THE CRAWFORDS' BIG BOOK OF Math-tivities

THE CRAWFORDS' BIG BOOK OF Math-tivities

BRIAN & YVONNE CRAWFORD

COMPASS

The Crawfords' BIG Book of Math-tivities

Copyright © 2013 by Brian Crawford and Yvonne Crawford

All rights reserved.

Your individual purchase of this book entitles you to reproduce these pages as needed for your own classroom use only. Otherwise, no part of this book may be reproduced or utilized in any way or by any means, electronic or mechanical, including photocopying, recording, or information storage or retrieval system, without prior written permission from the publisher. Individual copies may not be distributed in any other form.

Cover and book design by Jacob L. Grant
Original character artwork by Yvonne Crawford
Additional artwork by Jacob L. Grant

ISBN 978-1-9384062-9-4

To order multiple copies of the physical book or the digital download,
or for more information about other
Brigantine Media/Compass Publishing educational materials, visit
www.brigantinemedia.com/compass

Brigantine Media/Compass Publishing
211 North Avenue
St. Johnsbury, Vermont 05819
Phone: 802-751-8802
Fax: 802-751-8804
E-mail: **neil@brigantinemedia.com**

Dedication

For our families,
and for educators everywhere.

Contents

1 Introduction

3 Common Core Standards Alignment

13 **Chapter 1**
Mathbooking

37 **Chapter 2**
Math Glyphs

77 **Chapter 3**
Math and Tell

81 **Chapter 4**
Math Games and Puzzles

120 **Chapter 5**
Taking Math Outside

148 **Chapter 6**
Holiday and Seasonal Math

Introduction

If you teach young children that math involves fun and creativity, you can help them develop an early appreciation of mathematics. Students who enjoy math class will approach the subject with a high energy level and an open mind, making it easier for them to learn new skills. And if a student gains an early appreciation of math, that student will be more likely to succeed in mathematics in the future.

The aim of this book is to bring an element of fun to your math activities while teaching concepts that align with the Common Core State Standards for your grade level. In this book, you will find many new ways to mix the fun of arts, crafts, games, puzzles, and storytelling with learning math skills, all keyed to specific Standards of the Common Core. Using these methods, your students will learn new mathematical concepts in an entertaining and engaging way.

The book is divided into six chapters that introduce lots of "math-tivities" to help your students in kindergarten, first grade, and second grade learn mathematics. Some may be ideas you haven't seen before, while others are new twists on the kinds of activities you may be doing in other curricular areas. Chapters 1, 2, and 3 introduce new creative concepts for teaching math. Chapters 4, 5, and 6 offer a variety of activities that add to your repertoire of math teaching tools.

Chapter 1	Mathbooking
Chapter 2	Math Glyphs
Chapter 3	Math and Tell
Chapter 4	Math Games and Puzzles
Chapter 5	Taking Math Outside
Chapter 6	Holiday and Seasonal Math

A chart showing all the Common Core Standards that are aligned with the book's activities starts on **page 3**. All of the Common Core Standards for mathematics in kindergarten, grade 1, and grade 2 are included. Start with this chart every time you are ready to introduce a new math skill. There you'll find one or more math-tivities in the book that will help your students master that Standard.

There are plenty of ready-to-go math-tivities included in this book for you to print and use with your students. However, the math-tivities in the book are also meant to spur your own creativity in designing your own classroom activities. We include instructions and blank templates for many math-tivities to help you make your own.

We love math, art, puzzles, and games, so it was only natural to bring them all together in this BIG Book of Math-tivities!

Common Core Standards Alignment

Kindergarten..........................pages **4 – 6**

Grade 1..........................pages **7 – 9**

Grade 2..........................pages **10 – 12**

Kindergarten

Common Core Standards Alignment

Common Core Standards | Math-tivities

Counting & Cardinality	
K.CC.A.1 Count to 100 by ones and by tens.	Mathbooking prompt 1 (p. 22) Mathbooking prompt 3 (p. 23) Puppy Glyph (pp. 44-53) Color by Numbers (p. 93) Outdoor Mathbooking prompt 2 (p. 126) Counting Leaves (p. 145)
K.CC.A.2 Count forward beginning from a given number within the known sequence (instead of having to begin at 1).	Mathbooking prompt 8 (p. 25) Puppy Glyph (pp. 44-53)
K.CC.A.3 Write numbers from 0 to 20. Represent a number of objects with a written numeral 0-20 (with 0 representing a count of no objects).	Mathbooking prompt 9 (p. 26)
K.CC.B.4 Understand the relationship between numbers and quantities; connect counting to cardinality.	Mathbooking prompt 9 (p. 26) Counting Leaves (p. 145)
K.CC.B.5 Count to answer "how many?" questions about as many as 20 things arranged in a line, a rectangular array, or a circle, or as many as 10 things in a scattered configuration; given a number from 1–20, count out that many objects.	Mathbooking prompt 1 (p. 22) Mathbooking prompt 9 (p. 26) Outdoor Mathbooking prompt 1 (p. 126)
K.CC.C.6 Identify whether the number of objects in one group is greater than, less than, or equal to the number of objects in another group, e.g., by using matching and counting strategies.	Mathbooking prompt 5 (p. 24)
K.CC.C.7 Compare two numbers between 1 and 10 presented as written numerals.	Bubble Math (p. 96)
Operations & Algebraic Thinking	
K.OA.A.1 Represent addition and subtraction with objects, fingers, mental images, drawings, sounds (e.g., claps), acting out situations, verbal explanations, expressions, or equations.	Mathbooking prompt 2 (p. 22) Mathbooking prompt 4 (p. 23) Mathbooking prompt 9 (p. 26) Math Libs (p. 90)
K.OA.A.2 Solve addition and subtraction word problems, and add and subtract within 10, e.g., by using objects or drawings to represent the problem.	Harvest task card 8 (p. 155)
K.OA.A.3 Decompose numbers less than or equal to 10 into pairs in more than one way, e.g., by using objects or drawings, and record each decomposition by a drawing or equation (e.g., 5 = 2 + 3 and 5 = 4 + 1).	Harvest task card 5 (p. 154)

Common Core Standards Alignment  Kindergarten

Common Core Standards | Math-tivities

Common Core Standards	Math-tivities
K.OA.A.4 For any number from 1 to 9, find the number that makes 10 when added to the given number, e.g., by using objects or drawings, and record the answer with a drawing or equation.	Harvest task card 6 (p. 154)
K.OA.A.5 Fluently add and subtract within 5.	Mathbooking prompt 2 (p. 22) Puppy Glyph (pp. 44-53) Garden Find and Draw (pp. 100-102) Number Search (p. 109) Match the Math (pp. 112-114) Hide, Seek, and Solve (pp. 133-135)
Number & Operations in Base Ten	
K.NBT.A.1 Compose and decompose numbers from 11 to 19 into ten ones and some further ones, e.g., by using objects or drawings, and record each composition or decomposition by a drawing or equation (such as 18 = 10 + 8); understand that these numbers are composed of ten ones and one, two, three, four, five, six, seven, eight, or nine ones.	Harvest task card 2 (p. 153)
Measurement & Data	
K.MD.A.1 Describe measurable attributes of objects, such as length or weight. Describe several measurable attributes of a single object.	Outdoor Mathbooking prompt 3 (p. 127)
K.MD.A.2 Directly compare two objects with a measurable attribute in common, to see which object has "more of"/"less of" the attribute, and describe the difference. For example, directly compare the heights of two children and describe one child as taller/shorter.	Mathbooking prompt 6 (p. 24) Harvest task card 7 (p. 154)
K.MD.B.3 Classify objects into given categories; count the numbers of objects in each category and sort the categories by count.	Outdoor Mathbooking prompt 4 (p. 127) Counting Leaves (p. 145)
Geometry	
K.G.A.1 Describe objects in the environment using names of shapes, and describe the relative positions of these objects using terms such as above, below, beside, in front of, behind, and next to.	Mathbooking prompt 7 (p. 25) Harvest task card 9 (p. 155)
K.G.A.2 Correctly name shapes regardless of their orientations or overall size.	Mathbooking prompt 10 (p. 26)
K.G.A.3 Identify shapes as two-dimensional (lying in a plane, "flat") or three-dimensional ("solid").	Harvest task card 4 (p. 154)

Kindergarten  Common Core Standards Alignment

Common Core Standards Math-tivities

Common Core Standards	Math-tivities
K.G.B.4 Analyze and compare two- and three-dimensional shapes, in different sizes and orientations, using informal language to describe their similarities, differences, parts (e.g., number of sides and vertices/"corners") and other attributes (e.g., having sides of equal length).	Harvest task card 3 (p. 153)
K.G.B.5 Model shapes in the world by building shapes from components (e.g., sticks and clay balls) and drawing shapes.	Harvest task card 1 (p. 153)
K.G.B.6 Compose simple shapes to form larger shapes. For example, "Can you join these two triangles with full sides touching to make a rectangle?"	Harvest task card 10 (p. 155)

Common Core Standards Alignment  Grade 1

Common Core Standards — Math-tivities

Common Core Standards	Math-tivities
Operations & Algebraic Thinking	
1.OA.A.1 Use addition and subtraction within 20 to solve word problems involving situations of adding to, taking from, putting together, taking apart, and comparing, with unknowns in all positions, e.g., by using objects, drawings, and equations with a symbol for the unknown number to represent the problem.	Mathbooking prompt 7 (p. 30)
1.OA.A.2 Solve word problems that call for addition of three whole numbers whose sum is less than or equal to 20, e.g., by using objects, drawings, and equations with a symbol for the unknown number to represent the problem.	Mathbooking prompt 2 (p.27)
1.OA.B.3 Apply properties of operations as strategies to add and subtract.2 Examples: If 8 + 3 = 11 is known, then 3 + 8 = 11 is also known. (Commutative property of addition.) To add 2 + 6 + 4, the second two numbers can be added to make a ten, so 2 + 6 + 4 = 2 + 10 = 12. (Associative property of addition.)	Winter task card 7 (p. 158)
1.OA.B.4 Understand subtraction as an unknown-addend problem. For example, subtract 10 – 8 by finding the number that makes 10 when added to 8.	Winter task card 3 (p. 157)
1.OA.C.5 Relate counting to addition and subtraction (e.g., by counting on 2 to add 2).	Winter task card 4 (p. 158)
1.OA.C.6 Add and subtract within 20, demonstrating fluency for addition and subtraction within 10. Use strategies such as counting on; making ten (e.g., 8 + 6 = 8 + 2 + 4 = 10 + 4 = 14); decomposing a number leading to a ten (e.g., 13 – 4 = 13 – 3 – 1 = 10 – 1 = 9); using the relationship between addition and subtraction (e.g., knowing that 8 + 4 = 12, one knows 12 – 8 = 4); and creating equivalent but easier or known sums (e.g., adding 6 + 7 by creating the known equivalent 6 + 6 + 1 = 12 + 1 = 13).	Mathbooking prompt 9 (p. 31) Puppy Glyph (pp. 54-63) Math Libs (p. 91) Color by Addition (p. 94) Pet Find and Draw (pp. 103-105) Number Search (p. 110) Match the Math (pp. 115-117) Outdoor Mathbooking prompt 2 (p. 128) Outdoor Mathbooking prompt 3 (p. 129) Hide, Seek, and Solve (pp. 137-139)
1.OA.D.7 Understand the meaning of the equal sign, and determine if equations involving addition and subtraction are true or false. For example, which of the following equations are true and which are false? 6 = 6, 7 = 8 – 1, 5 + 2 = 2 + 5, 4 + 1 = 5 + 2.	Winter task card 5 (p. 158)

Grade 1 — Common Core Standards Alignment

Common Core Standards | Math-tivities

Common Core Standards	Math-tivities
1.OA.D.8 Determine the unknown whole number in an addition or subtraction equation relating three whole numbers. For example, determine the unknown number that makes the equation true in each of the equations 8 + ? = 11, 5 = _ – 3, 6 + 6 = _.	Winter task card 6 (p. 158)

Number & Operations in Base Ten

Common Core Standards	Math-tivities
1.NBT.A.1 Count to 120, starting at any number less than 120. In this range, read and write numerals and represent a number of objects with a written numeral.	Outdoor Mathbooking prompt 4 (p. 129) Winter task card 4 (p. 158)
1.NBT.B.2 Understand that the two digits of a two-digit number represent amounts of tens and ones.	Mathbooking prompt 10 (p. 31)
1.NBT.B.3 Compare two two-digit numbers based on meanings of the tens and ones digits, recording the results of comparisons with the symbols >, =, and <.	Bubble Math (p. 97)
1.NBT.C.4 Add within 100, including adding a two-digit number and a one-digit number, and adding a two-digit number and a multiple of 10, using concrete models or drawings and strategies based on place value, properties of operations, and/or the relationship between addition and subtraction; relate the strategy to a written method and explain the reasoning used. Understand that in adding two-digit numbers, one adds tens and tens, ones and ones; and sometimes it is necessary to compose a ten.	Winter task card 9 (p. 159)
1.NBT.C.5 Given a two-digit number, mentally find 10 more or 10 less than the number, without having to count; explain the reasoning used.	Winter task card 1 (p. 157)
1.NBT.C.6 Subtract multiples of 10 in the range 10-90 from multiples of 10 in the range 10-90 (positive or zero differences), using concrete models or drawings and strategies based on place value, properties of operations, and/or the relationship between addition and subtraction; relate the strategy to a written method and explain the reasoning used.	Winter task card 1 (p. 157)

Measurement & Data

Common Core Standards	Math-tivities
1.MD.A.1 Order three objects by length; compare the lengths of two objects indirectly by using a third object.	Mathbooking prompt 3 (p. 28) Mathbooking prompt 5 (p. 29) Hide, Seek, and Solve (pp. 137-139) Comparing Leaves (p. 146)

Common Core Standards Alignment  Grade 1

Common Core Standards | Math-tivities

Common Core Standards	Math-tivities
1.MD.A.2 Express the length of an object as a whole number of length units, by laying multiple copies of a shorter object (the length unit) end to end; understand that the length measurement of an object is the number of same-size length units that span it with no gaps or overlaps. Limit to contexts where the object being measured is spanned by a whole number of length units with no gaps or overlaps.	Winter task card 2 (p. 157)
1.MD.B.3 Tell and write time in hours and half-hours using analog and digital clocks.	Mathbooking prompt 1 (p. 27) Hide, Seek, and Solve (pp. 137-139) Outdoor Mathbooking prompt 1 (p. 128)
1.MD.C.4 Organize, represent, and interpret data with up to three categories; ask and answer questions about the total number of data points, how many in each category, and how many more or less are in one category than in another.	Winter task card 8 (p. 159)
Geometry	
1.G.A.1 Distinguish between defining attributes (e.g., triangles are closed and three-sided) versus non-defining attributes (e.g., color, orientation, overall size); build and draw shapes to possess defining attributes.	Mathbooking prompt 8 (p. 30)
1.G.A.2 Compose two-dimensional shapes (rectangles, squares, trapezoids, triangles, half-circles, and quarter-circles) or three-dimensional shapes (cubes, right rectangular prisms, right circular cones, and right circular cylinders) to create a composite shape, and compose new shapes from the composite shape.	Winter task card 10 (p. 159)
1.G.A.3 Partition circles and rectangles into two and four equal shares, describe the shares using the words halves, fourths, and quarters, and use the phrases half of, fourth of, and quarter of. Describe the whole as two of, or four of the shares. Understand for these examples that decomposing into more equal shares creates smaller shares.	Mathbooking prompt 4 (p. 28) Mathbooking prompt 6 (p. 29)

Grade 2

Common Core Standards Alignment

Common Core Standards | Math-tivities

Operations & Algebraic Thinking

Common Core Standards	Math-tivities
2.OA.A.1 Use addition and subtraction within 100 to solve one- and two-step word problems involving situations of adding to, taking from, putting together, taking apart, and comparing, with unknowns in all positions, e.g., by using drawings and equations with a symbol for the unknown number to represent the problem.	Mathbooking prompt 7 (p. 35)
2.OA.B.2 Fluently add and subtract within 20 using mental strategies. By end of Grade 2, know from memory all sums of two one-digit numbers.	Puppy Glyph (pp. 64-73)
2.OA.C.3 Determine whether a group of objects (up to 20) has an odd or even number of members, e.g., by pairing objects or counting them by 2s; write an equation to express an even number as a sum of two equal addends.	Mathbooking prompt 3 (p. 33)
2.OA.C.4 Use addition to find the total number of objects arranged in rectangular arrays with up to 5 rows and up to 5 columns; write an equation to express the total as a sum of equal addends.	Spring task card 1 (p. 161)

Numbers & Operations in Base Ten

Common Core Standards	Math-tivities
2.NBT.A.1 Understand that the three digits of a three-digit number represent amounts of hundreds, tens, and ones; e.g., 706 equals 7 hundreds, 0 tens, and 6 ones. Understand the following as special cases:	Spring task card 2 (p. 161)
2.NBT.A.2 Count within 1000; skip-count by 5s, 10s, and 100s.	Mathbooking prompt 9 (p. 36) Outdoor Mathbooking prompt 2 (p. 130)
2.NBT.A.3 Read and write numbers to 1000 using base-ten numerals, number names, and expanded form.	Mathbooking prompt 10 (p. 36)
2.NBT.A.4 Compare two three-digit numbers based on meanings of the hundreds, tens, and ones digits, using >, =, and < symbols to record the results of comparisons.	Bubble Math (p. 98)
2.NBT.B.5 Fluently add and subtract within 100 using strategies based on place value, properties of operations, and/or the relationship between addition and subtraction.	Puppy Glyph (pp. 64-73) Math Libs (p. 92) Color by Addition (p. 95) School Find and Draw (pp. 106-108) Number Search (p. 111) Hide, Seek, and Solve (pp. 141-143)
2.NBT.B.6 Add up to four two-digit numbers using strategies based on place value and properties of operations.	Spring task card 3 (p. 161)

Common Core Standards Alignment

Common Core Standards — Math-tivities

Common Core Standards	Math-tivities
2.NBT.B.7 Add and subtract within 1000, using concrete models or drawings and strategies based on place value, properties of operations, and/or the relationship between addition and subtraction; relate the strategy to a written method. Understand that in adding or subtracting three-digit numbers, one adds or subtracts hundreds and hundreds, tens and tens, ones and ones; and sometimes it is necessary to compose or decompose tens or hundreds.	Spring task card 10 (p. 163)
2.NBT.B.8 Mentally add 10 or 100 to a given number 100–900, and mentally subtract 10 or 100 from a given number 100–900.	Mathbooking prompt 10 (p. 36)
2.NBT.B.9 Explain why addition and subtraction strategies work, using place value and the properties of operations.	Mathbooking prompt 8 (p. 35)
Measurement & Data	
2.MD.A.1 Measure the length of an object by selecting and using appropriate tools such as rulers, yardsticks, meter sticks, and measuring tapes.	Mathbooking prompt 1 (p. 32) Outdoor Mathbooking prompt 3 (p. 131) Mathbooking prompt 4 (p. 33)
2.MD.A.2 Measure the length of an object twice, using length units of different lengths for the two measurements; describe how the two measurements relate to the size of the unit chosen.	Mathbooking prompt 4 (p. 33)
2.MD.A.3 Estimate lengths using units of inches, feet, centimeters, and meters.	Mathbooking prompt 4 (p. 33) Outdoor Mathbooking prompt 4 (p. 131) Estimating Leaf Length (p. 147)
2.MD.A.4 Measure to determine how much longer one object is than another, expressing the length difference in terms of a standard length unit.	Mathbooking prompt 1 (p. 32)
2.MD.B.5 Use addition and subtraction within 100 to solve word problems involving lengths that are given in the same units, e.g., by using drawings (such as drawings of rulers) and equations with a symbol for the unknown number to represent the problem.	Mathbooking prompt 6 (p. 34)
2.MD.B.6 Represent whole numbers as lengths from 0 on a number line diagram with equally spaced points corresponding to the numbers 0, 1, 2, ..., and represent whole-number sums and differences within 100 on a number line diagram.	Spring task card 7 (p. 162)
2.MD.C.7 Tell and write time from analog and digital clocks to the nearest five minutes, using a.m. and p.m.	Mathbooking prompt 2 (p. 32)

Grade 2 Common Core Standards Alignment

Common Core Standards | **Math-tivities**

Common Core Standards	Math-tivities
2.MD.C.8 Solve word problems involving dollar bills, quarters, dimes, nickels, and pennies, using $ and ¢ symbols appropriately. Example: If you have 2 dimes and 3 pennies, how many cents do you have?	Mathbooking prompt 5 (p. 34) Puppy Glyph (p. 64-73)
2.MD.D.9 Generate measurement data by measuring lengths of several objects to the nearest whole unit, or by making repeated measurements of the same object. Show the measurements by making a line plot, where the horizontal scale is marked off in whole-number units.	Spring task card 9 (p. 163)
2.MD.D.10 Draw a picture graph and a bar graph (with single-unit scale) to represent a data set with up to four categories. Solve simple put-together, take-apart, and compare problems using information presented in a bar graph.	Spring task card 6 (p. 162)
Geometry	
2.G.A.1 Recognize and draw shapes having specified attributes, such as a given number of angles or a given number of equal faces. Identify triangles, quadrilaterals, pentagons, hexagons, and cubes.	Spring task card 4 (p. 162)
2.G.A.2 Partition a rectangle into rows and columns of same-size squares and count to find the total number of them.	Spring task card 8 (p. 163)
2.G.A.3 Partition circles and rectangles into two, three, or four equal shares, describe the shares using the words halves, thirds, half of, a third of, etc., and describe the whole as two halves, three thirds, four fourths. Recognize that equal shares of identical wholes need not have the same shape.	Spring task card 5 (p. 162)

Chapter 1

Math Journals + Scrapbooking = Mathbooking

Mathbooking is a combination of math journaling and scrapbooking. It's a fun way for kids to use their creativity while learning about mathematics.

A traditional math journal is a student's notebook filled with daily math problems distributed by the teacher. Problems, called "prompts," are pasted into the math journal or copied into the journal from the board. Students answer the prompts in their math journals. Math journals are a useful tool for doing math every day.

But they are **boring**!

Mathbooking is like traditional math journaling—but kicked up a few notches:

Traditional math journals	Mathbooks
Use a traditional notebook	**Use a scrapbook or photo album**
Feature printed prompts or prompts copied from the board	**Feature colorful, creative prompts**
Questions can be used at any time, year after year	**Questions are fresh, timely, and tie into a unit, holiday, or season**
Answers are in the form of numbers	**Answers can be in the form of numbers, words, drawings, patterns, photos, and more**
Creative doodling in the math journal is frowned upon	**Creative doodling in the Mathbook is considered part of the learning process**

Mathbooking pumps energy into your regular math class. Kids solve math problems while letting their creative juices flow! Your students will learn to equate *mathematics* with *fun*!

Mathbooking Topics

Mathbooking topics are almost limitless; the key is to promote understanding of a variety of mathematical skills by tackling different kinds of math problems throughout a school year. One day, a prompt might test a student's skills with measurement and data analysis, to reinforce learning of Common Core Standard 2.MD.A.1:

 Look at this funny clown shoe. What would you use to measure it? Measure the clown shoe. Next, draw a picture of your own shoe in your Mathbook and measure it. Which shoe is bigger?

The next day students might tackle a Mathbook prompt that covers Common Core Standard 2.MD.C.7 to reinforce learning to tell time from analog and digital clocks to the nearest five minutes:

 It's back to school time, and Isabella has to wake up at 6:30 a.m. In the picture of the clock, draw the time that Isabella has to wake up. Then draw your own clock and draw the time you wake up each morning for school.

Using Mathbook prompts with diverse topics based on material you have taught throughout the school year gives your students a great review and engages them to learn and have fun during math class.

Mathbooking Uses

A Mathbook is a very adaptable teaching tool. Once you try Mathbooking, you'll find more and more ways to incorporate it into your math classes. Here are a number of uses for Mathbooking:

- **Homework**

    At the end of your math class, hand out Mathbook prompts for students to paste into their Mathbooks and complete for homework. Depending on how much homework you want to give your students, you can provide one or more prompts of varying difficulty for them to complete. Make sure that you take time to explain the problems so that they don't have any trouble understanding the prompt at home. The next day, you can go over the Mathbooks together as a class, sharing the colorful answers provided by the students. Some teachers prefer to flip through students' Mathbooks every few days to track their progress.

> **TIP**
>
> As a homework assignment, give your students a Mathbook prompt that involves going outside. For example, have them measure leaves, count rocks, or estimate how long it will take them to run the length of the driveway.

✦ For early finishers

Some students are quicker than others, and you're likely to have some early finishers. You can use Mathbooks to help your advanced students learn more skills in mathematics and keep them working while other students are completing their work more slowly. If the prompts are fun to complete, your advanced students will feel that you are rewarding them for their hard work and quick understanding.

✦ Part of a reward system

Use Mathbooks in conjunction with a reward system for hard work or homework completion. For example, for every 20 Mathbook prompts completed by a student, you could allow one trip to the treasure box, an extra book loan from the classroom library, or any other reward system that you use in your classroom.

> **TIP**
>
> If you're rewarding students for Mathbooking, provide rewards that are related to Mathbooking. For jobs well done, you can give your students extra clip art, metallic star stickers, funky-colored crayons, or tubes of glitter that they can use while Mathbooking.

✦ Exit slips

Mathbook prompts can be used as exit slips. If recess is approaching, hand out Mathbook prompts to be completed and turned in before the students head outside to play.

✦ Holidays and special occasions

Holidays are great times to use Mathbooks. There are plenty of holidays and special days throughout the year to get excited about—Halloween, Mother's Day, Easter, President's Day—the list goes on!

Don't be limited by the usual holidays. If it's your town's 100th birthday, create a Mathbook prompt to tie into your town's celebration. Create a Mathbook prompt when your school serves its 5,000th meal. Turn to the Mathbooks when the teacher's dog wins a ribbon in the local pet show . . . well, you get the idea.

- **To demonstrate work to parents**
When you're showing parents their children's progress in mathematics over the school year, Mathbooks are a powerful and comprehensive presentation tool. They showcase the analytical and problem-solving skills that your students have practiced throughout the school year and demonstrate that you and your students have a good time while learning.

Mathbooking and the Common Core

Mathbooking offers many ways to teach and review the Common Core Standards for mathematics:

- **Provide prompts as a Common Core review for previously taught standards**
At the beginning of the school year, use Mathbook prompts covering Common Core Standards for your students' previous year to gauge the knowledge of your incoming students. This will help to guide your lessons throughout the first few weeks.

- **Provide prompts for every Common Core Standard for your grade**
Create Mathbook prompts for every skill in the Common Core State Standards for your grade level throughout the school year. You can gauge how well your students have assimilated the Common Core material by flipping through their Mathbooks to see the responses that they have given to the prompts for specific Common Core Standards.

- **Provide prompts for areas where students need extra attention**
If a student is falling behind on learning a certain Common Core Standard, you can provide extra Mathbook prompts that focus on that Common Core skill and offer more opportunities for practice.

TIP

The Common Core Alignment chart in this book (**pages 3-12**) is a great tool to help track how you have incorporated all the Standards for your grade into your students' Mathbook prompts.

Benefits

The more you use Mathbooking, the easier it is to see its benefits:

- **Adds versatility to your teaching**
Mathbooking can introduce, practice, or review skills as needed.

- **Provides a transitional activity**
Mathbooking is a perfect activity to assign when you're shifting to a new subject.
- **Gives structure and familiarity**
When used daily, weekly, or on a regular basis, Mathbooking is part of an orderly classroom schedule.
- **Takes place over time**
It's no fun to open up a scrapbook to find only one page of pictures pasted inside. In the same way, it's no fun to do Mathbooking one day and never participate in the activity again. View Mathbooking as a journey; every day your students will be learning new math skills, then using those skills in their Mathbooks.
- **Offers feedback on students' progress**
A full Mathbook has a wealth of information about how well a student has learned math concepts throughout the year.

Getting Started

When first starting to use Mathbooks in your classroom, you'll have to make some choices:

1 What type of book should my students use to create a Mathbook?
- Scrapbook (like those from a hobby or craft store)
- Photo album
- Loose-leaf, bound, or spiral notebook

2 How often do I want students to use their Mathbooks?
- Once a day
- Three times a week
- Once a week
- Other

3 When do I want my students to use their Mathbooks?
- At the beginning of class
- During class
- At the end of class
- For homework assignments
- As a treat during math class
- Any time the mood strikes us

4 **What kinds of questions are my students going to solve while Mathbooking?**
- Review questions
- Questions to introduce a new topic
- Questions where answering problems in a visual manner is helpful to assimilate learning

Decide what works best for you in your own classroom. Any combination of the choices given will work.

Once you have decided what role Mathbooking will play in your classroom, introduce it to your class. Make sure your class understands that Mathbooking will be fun—a creative, no-holds-barred approach to tackling math problems in an entertaining, visual manner. Allow your students to be as creative as they wish when answering the prompts. It will take a little time and effort to make Mathbooks a consistent part of your daily schedule, but in the end, Mathbooks will leave your students with the positive message that solving math problems is a lot of fun.

Mathbooking Procedure

Here are specific instructions for how to do Mathbooking. It's fun and easy:

Supplies:
- Mathbook prompt
- Scissors
- Blank Mathbook page or Mathbooks for each student
- Paste or glue
- Crayons or colored pencils
- Other art supplies

Instructions:
- Photocopy and cut out the Mathbook prompts before class so the prompts are ready for your students to use. Cut out and distribute one prompt per student.
- Have students paste or glue the prompt into their Mathbooks.
- Students solve the math problem in the prompt and show their work in their Mathbooks.

> **TIP**
>
> Have your students tape a ribbon to the cover of their Mathbooks. When they work on a Mathbook page, they can slip the other end of the ribbon onto that page before closing their Mathbooks. That way, they will easily be able to find where they are in their Mathbooks the next time they have a prompt to solve.

Creating Mathbook Prompts

Ten complete Mathbook prompts for each grade K - 2 are included at the end of this chapter. But since the best Mathbook prompts relate to your specific curriculum, you will want to create new prompts for topics as they arise. Here are some tips to get you started:

- **Make your prompts timely**

   If the next holiday coming up in your calendar is Halloween, create Mathbook prompts with a Halloween theme for your students to tackle. Coming up with timely prompts will help to keep your students engaged and excited about Mathbooking. Use your creativity to find subject matter: ghosts, goblins, witches, vampires . . . when it comes to making creative Mathbook prompts, anything goes! Just keep in mind the reading and comprehension level of your students.

- **Model creativity in your prompts**

   You want your students to think creatively when they are Mathbooking, so you'll need to use your creativity when you develop Mathbook prompts. You will be printing out the prompts for each student to paste in his/her Mathbook. By using a few different fonts and some fun clip art or your own drawings, you will model the creative approach you want your students to take with their responses. Employ italics and bold lettering to help draw students' attention to important information in the Mathbook prompts you create.

   Students who answer Mathbook prompts may come up with a variety of different answers depending on their skill levels and how creative they choose to be. Your goal as a teacher is to make sure that your students understand the mathematical skills that accompany each problem while also encouraging their creativity.

- **Create different types of math problems**

   Variety, as they say, is the spice of life, and Mathbooking is a great way to spice up your math lessons. Your Mathbook prompts should challenge your students with different types of math problems for them to tackle.

   When creating Mathbook prompts, it is easy to fall into the habit of creating simple math equations, such as addition or subtraction problems, for your students to solve with a pencil. But to fully engage your students, dig deeper into your toolbox of skills. Find ways for students to use their whole body to solve math problems—measuring items, counting their steps, or exploring their environment for answers. Make your Mathbook prompts reflect the more creative nature of Mathbooking itself, so that students learn to develop innate mathematical knowledge to solve problems.

Almost any kind of math problem can be used to inspire a Mathbook prompt:

- Word problems
- Graphs and charts
- Tables
- Basic operations (addition, subtraction, multiplication, division)
- Fact families
- Comparison problems
- Patterns
- Ordering
- Spatial problems involving grids or timelines
- Skip counting problems
- Greater than and less than problems

Your prompts will be based on your curriculum, the specific material you are working on, and if applicable, the Common Core Standards for your grade.

Solving Mathbook Prompts

The Mathbook prompt should guide the student toward a method to answer the question, whether the answer is a simple number response or a more complex drawing that describes the operation needed to solve the problem. Make sure the prompt is clear, so students understand exactly what is expected to constitute a complete answer.

Some students—many kindergarteners, for example—may not be able to read the Mathbook prompts they are given. In this case, present the Mathbook prompts to your class on the board and demonstrate the problem. Then they can address the prompt in their Mathbooks, solving the problem given using pictures and numbers.

Students may need assistance solving Mathbook prompts. While your students are working in their Mathbooks, move around the room to see what methods they are using to solve the problems and where they might need a nudge in the right direction. The best way to help students solve a Mathbook problem is to show them how to use the skill you are teaching without giving them the answer outright. For example, if a student is having trouble with a Mathbook prompt that requires the addition of two groups of apples, have the student draw the two groups of apples in the Mathbook, then count all of the items in the two groups. This gives them both the answer they're looking for and a method to solve similar problems in the future.

When students have answered their Mathbook prompts, you can work through the answers to the prompts as a group. Some Mathbook problems may have a simple answer: 2 apples + 2 apples = 4 apples, for example. Other problems might have multiple answers depending on your students. For example, a Mathbook prompt that asks students to measure the length of their forearm will not have the same answer for all students. For this type of question, choose students from your class to come up to the board to demonstrate how they solved the Mathbook prompt.

TIP

Students can store their Mathbooks in their desks, or you can create a special Mathbook shelf in your classroom. If you're using a Mathbook shelf, print the students' names on the spines of their Mathbooks so that the Mathbooks will be easy to find during your next math class.

Remember that creativity is key in responding to Mathbook prompts; encourage your students to share the fun and imaginative ways that they answered their Mathbook prompts, and foster pride in the results they have created using mathematics.

Mathbook Prompts

Beginning on **page 22**, ten Mathbook prompts for each grade K – 2 are included, along with the corresponding Common Core Standard. Print these prompts and use them with your students when you are starting Mathbooking. They will serve as inspiration when you are ready to create your own Mathbook prompts.

Sophia went to the supermarket with her mother. She saw lots of apples. Circle the apples in groups of five and count how many apples Sophia found.

❶

CC Standards: K.CC.A.1; K.CC.B.5

Jacob and Mason went to the park. Jacob found 4 pebbles and Mason found 1 pebble. How many pebbles did they find?

Draw a picture of the pebbles to help you solve the problem.

❷

CC Standards: K.OA.A.1; K.OA.A.5

Isabella has a big box of pencils. Look at the numbers below and fill in the blanks with the missing numbers.
Draw five more pencils and continue the number pattern.

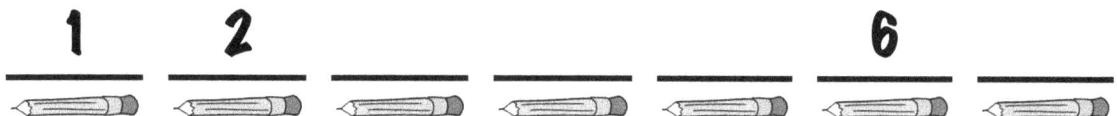

1   **2**   ___   ___   ___   **6**   ___

③   CC Standard: K.CC.A.1

Solve this problem by drawing pictures of suns to represent each number.

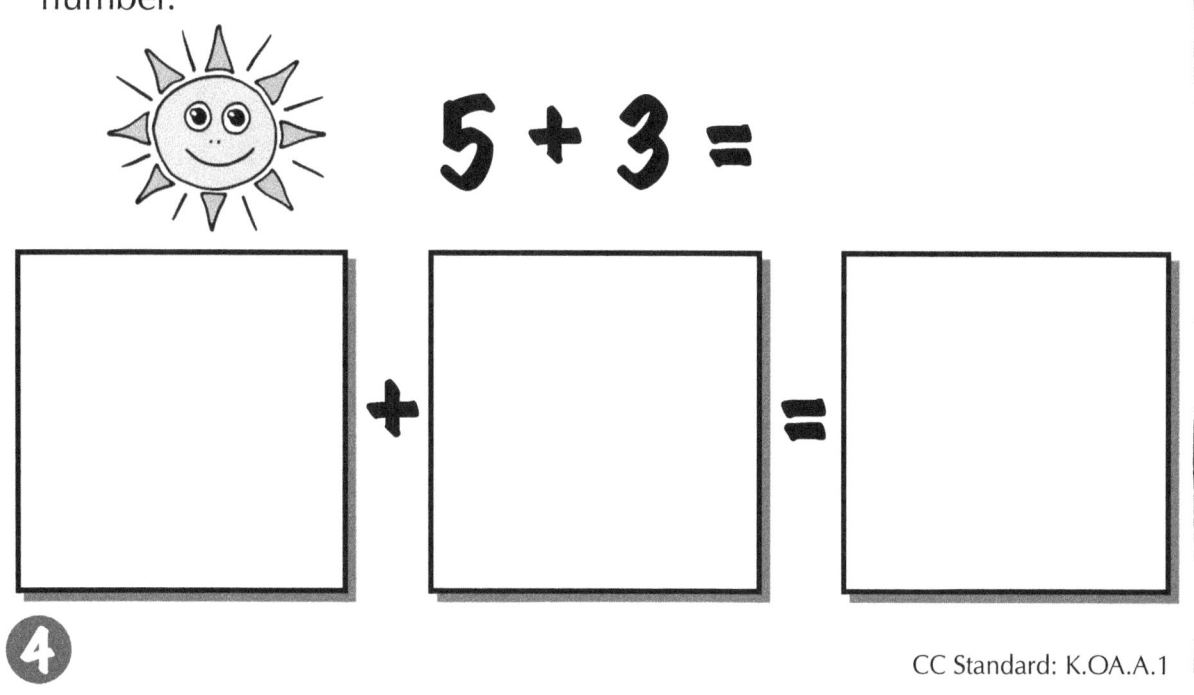

④   CC Standard: K.OA.A.1

Circle the group of kites with the most kites in it.

⑤

CC Standard: K.CC.C.6

Natalie is harvesting carrots with her Grandpa Joe. She harvested 3 carrots. Circle the carrot that is the longest.

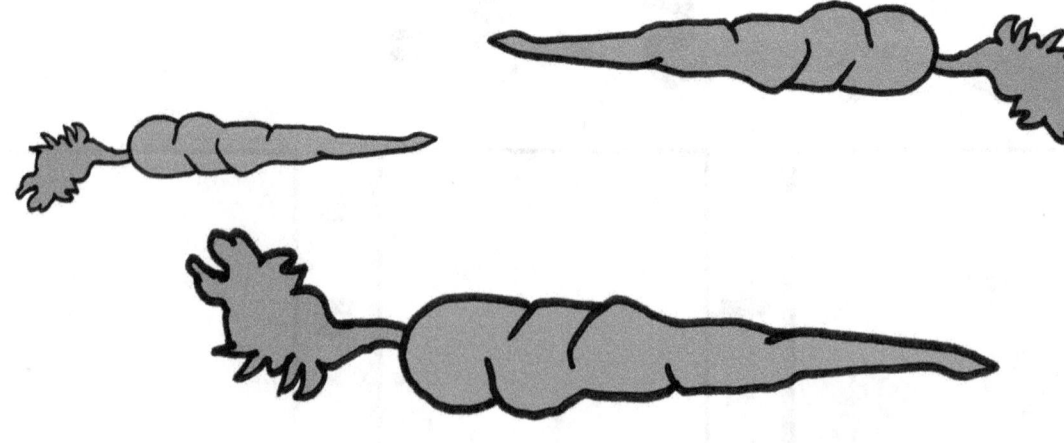

⑥

CC Standard: K.MD.A.2

Jacob is racing his car. Draw another race car behind his car. Then draw a sun above him. Last, draw a race track under his car.

CC Standard: K.G.A.1

Isabella is counting her pieces of candy. She knows she has 36 pieces of candy and she has already counted 8. Starting from 8, count to 36 to help Isabella finish counting.

CC Standard: K.CC.A.2

Draw 4 blue crayons. Next, draw 3 yellow crayons. How many crayons did you draw in all?

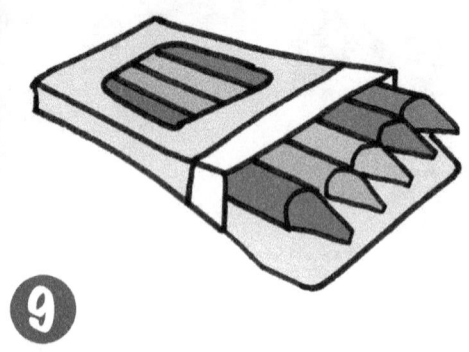

⑨

CC Standards: K.CC.A.3, K.CC.B.4, K.CC.B.5, K.OA.A.1

Isabella is learning how to draw circles and squares. Draw both a circle and a square so that she can see how to draw them.

⑩

CC Standard: K.G.A.2

Natalie wanted to wake up early so that she could pick some flowers. She wanted to wake up at 7:00 a.m. Draw that time on the clock below. Next, draw your own clock and write the time that you wake up on it.

CC Standard: 1.MD.B.3

This cat is looking for some cat friends. He found 3 cat friends in front of his house and he found 6 cat friends behind his house. How many cat friends did he find in all?

CC Standard: 1.OA.A.2

David was playing in the garden. He found 3 grasshoppers. Circle the biggest grasshopper that he found.

3

CC Standard: 1.MD.A.1

Lily is eating her lunch and wants to share her apple with her best friend. Help her divide this apple into equal halves.

4

CC Standard: 1.G.A.3

Isabella and her family are having a barbecue. Take out your ruler and measure the corn cobs. Which corn cob is longer?
Draw another corn cob that is longer than these 2 corn cobs.

5

CC Standard: 1.MD.A.1

Yum, yum! Who wants pizza? Use your pencils to divide the pizzas into the parts listed below each pizza. Then draw your own pizza in your mathbook and divide it into fourths.

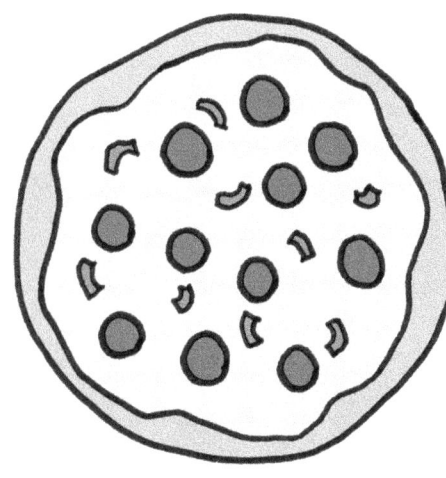 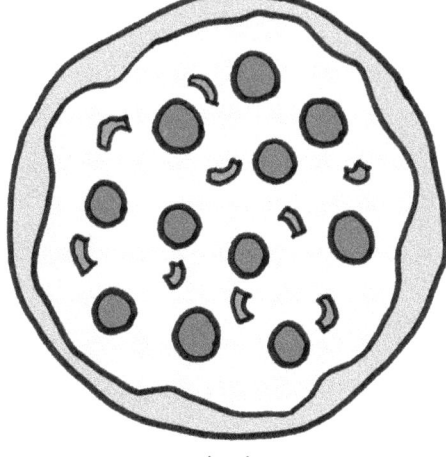

Halves   Thirds

6

CC Standard: 1.G.A.3

Emma has a pencil box that is 8 inches long. Aidan has a pencil box that is 6 inches long. Which pencil box is longer? How much longer is it?

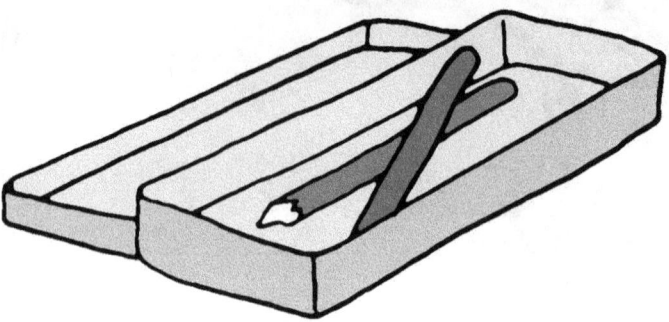

7

CC Standard: 1.OA.A.1

Lots of clocks are shaped like a circle.
Can you think of other things that are in the shape of a circle?
Draw as many things as you can see in your classroom that are in the shape of a circle.

8

CC Standard: 1.G.A.1

Mathbooking Prompts Grade 1

Mason loves math. It's his favorite subject in school! Today his teacher asked him to solve these equations. Can you help him?

8 + 4 =

1 + 10 =

12 + 2 =

5 + 5 =

9

CC Standard: 1.OA.C.6

Lily brought juice boxes for her classmates at school. How many juice boxes did she bring in all?

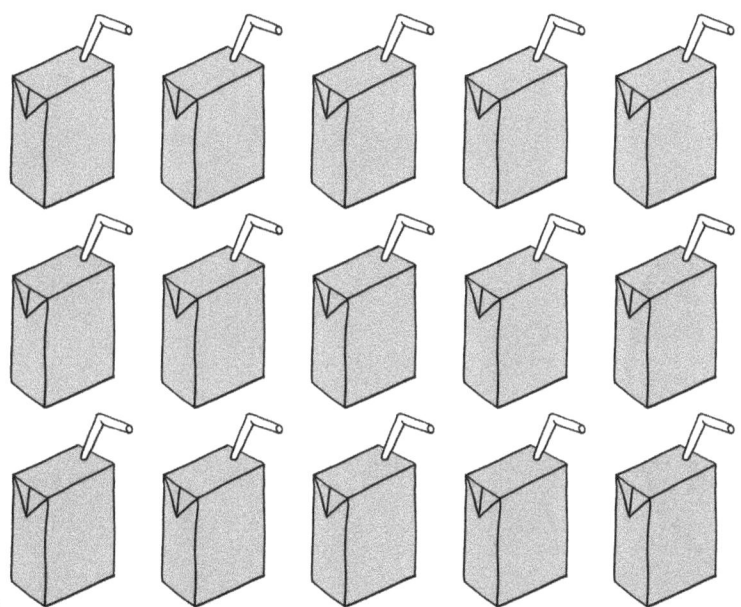

_____   _____
Groups   Ones
of ten

10

CC Standard: 1.NBT.B.2

Look at this funny clown shoe. What would you use to measure it? Measure the clown shoe. Next, draw a picture of your own shoe in your Mathbook and measure it. Which shoe is bigger?

❶

CC Standards: 2.MD.A.1, 2.MD.A.4

It's back to school time, and Isabella has to wake up at 6:30 a.m. to catch the bus. In the picture of the clock, draw the time that Isabella wakes up. Then draw your own clock and draw the time you wake up each morning for school.

❷

CC Standard: 2.MD.C.7

Mathbooking Prompts Grade 2

Squire Squirrel had a great afternoon collecting acorns. Did he find an even or odd number of acorns in all?

③

CC Standard: 2.OA.C.3

Look at these tools. Estimate the length of each tool in inches. Then measure each tool in both inches and centimeters with your ruler to see how close you were with your estimates.

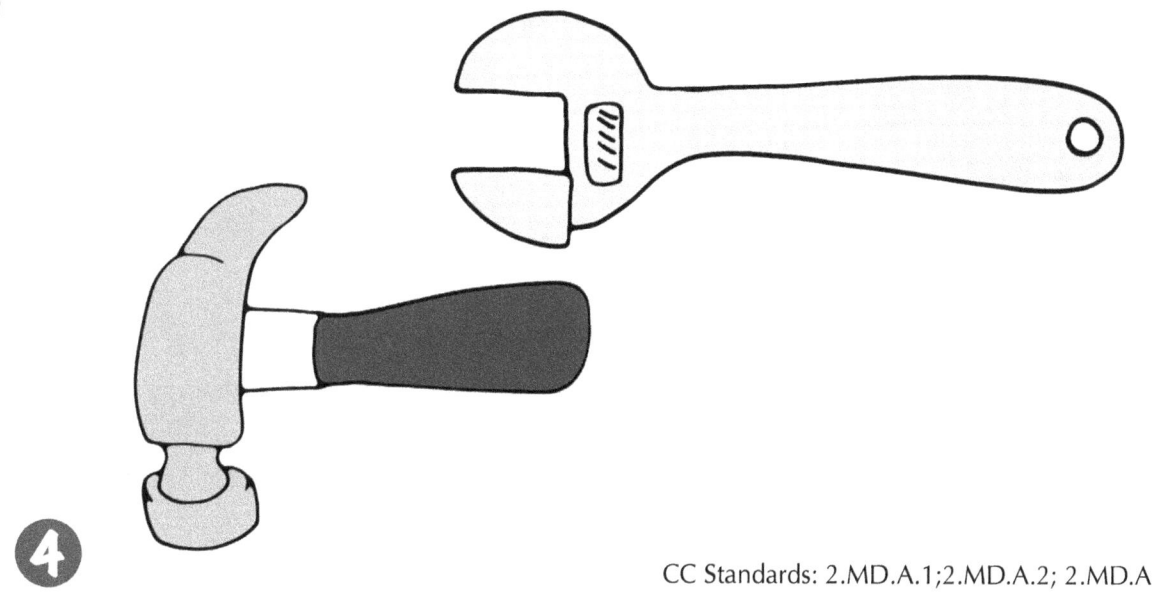

④

CC Standards: 2.MD.A.1; 2.MD.A.2; 2.MD.A.3

Brian & Yvonne Crawford ✱ ***Chapter 1*** ✱ *The Crawfords' BIG Book of Math-tivities*

Lily wants to buy some ice cream. She has 2 dimes and 4 nickels. How much money does Lily have? Draw pictures of the money to help solve the problem.

5

CC Standard: 2.MD.C.8

David found two ripe apples on the tree in his backyard. One apple was 8 inches tall. The second apple was 3 inches tall. What is the size difference between the two apples?

6

CC Standard: 2.MD.B.5

Sophia is helping her mother plant flowers. She planted 25 blue flowers in the backyard and 38 orange flowers out front. How many flowers did she plant in all?

CC Standard: 2.OA.A.1

Aidan is new to Victoria Elementary school. He is learning about related facts, but doesn't yet understand what they are. Can you tell him a related subtraction fact for **8 + 2 = 10**? Explain why the facts are related.

CC Standard: 2.NBT.B.9

David and his aunt are making cookies for school. They need to make 100 cookies for the entire second grade class. Draw groups of ten cookies to help you count to 100 by tens.

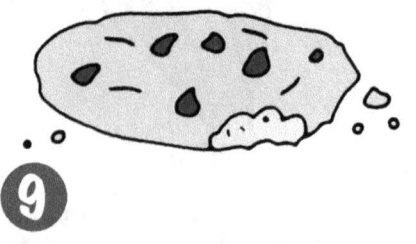

CC Standard: 2.NBT.A.2

Emma is picking strawberries. Emma picked 57 strawberries and her little sister picked 24 strawberries. How many strawberries did they pick in all? Write that number in words.

CC Standards: 2.NBT.A.3, 2.NBT.B.8

Chapter 2

Math Glyphs

Question + Math Problem + Artwork = Math Glyph

Glyphs are crafts and drawings that are shaped by each student's personality or interests, based on the way he/she answers a series of multiple choice questions. Each student's glyph is slightly different from the other students' glyphs, which is part of the fun!

Here's how a glyph works: Students answer a series of questions, each with a limited number of multiple-choice answers. After answering each question, the student is directed to craft his or her glyph in a certain manner.

For example: If a glyph involves drawing a figure, one question might be:

What is your favorite color?
- If your favorite color is blue, draw stars on your figure's shirt.
- If your favorite color is green, draw stripes on your figure's shirt.
- If your favorite color is yellow, draw spirals on your figure's shirt.
- If your favorite color is pink, draw zigzags on your figure's shirt.
- If your favorite color is something else, draw polka dots on your figure's shirt.

Other questions might direct you to color your figure's eyes a certain hue based on your favorite food; add shorts, pants, or overalls to your figure based on what time you go to bed; put a book, baseball, or paintbrush in your figure's hand based on the season you were born; and style your figure's hair short, medium, or long based on your favorite book. The students in every classroom have a variety of ages, birthdays, and interests, so each student's glyph is unique.

Math Glyphs

Making math glyphs involves adding a math problem to each personal response. The glyph directions are keyed from the answers to the math problems. As an example, using the same question from the general-purpose glyph (above), a math glyph adds a math problem for the students to solve, which then gives the directions for creating the glyph.

For example: If a math glyph involves creating a figure, one question might be:

What is your favorite color?
Answer one of the following math problems based on your favorite color:
- If your favorite color is blue: 4 + 4 = _____
- If your favorite color is green: 3 + 2 = _____
- If your favorite color is yellow: 5 + 2 = _____
- If your favorite color is pink: 2 + 2 = _____
- If your favorite color is something else: 2 + 7 = _____

Now add a pattern to your figure's shirt depending on your answer:
- If your answer is 7, draw purple stars on your figure's shirt.
- If your answer is 9, draw yellow stripes on your figure's shirt.
- If your answer is 8, draw orange spirals on your figure's shirt.
- If your answer is 4, draw black zigzags on your figure's shirt.
- If your answer is 5, draw brown polka-dots on your figure's shirt.

It's a good idea to design glyph questions in a way that every possible answer has an instruction. One answer can be crafted to work for answers not specifically covered, such as the "something else" answer in the math glyph example above.

Types of Math Glyphs

Math glyphs are limited only by your imagination. They can be simple, single-step creations to kick off a math lesson or complicated, multiple-step crafts that require an entire afternoon to complete.

Picture Math Glyphs

Students draw a picture based on the answers to multiple choice questions.

A picture math glyph can be a drawing of:

- A human figure, with clothes, hair, and eye color based on the glyph answers
- A pet, with type of animal, color, and accessories based on the glyph answers
- A landscape, with trees, fences, people, and pets based on the glyph answers
- A monster, with fur color, fangs, and number of eyes based on the glyph answers

If you can draw it, you can find a way to create a glyph from it!

Craft Math Glyphs

Students use scissors, glue, and construction paper to create craft math glyphs. They cut out shapes of varying sizes and colors and create collages with them based on their answers to math glyph questions. Another way to create craft glyphs is to find pictures in magazines or newspapers in response to glyph answers and create collages with those.

Sculpture Math Glyphs

Students use modeling clay to create sculptures based on answers to math questions. The sculptures can be figures, buildings, vehicles, monsters, dinosaurs, and more. Build a big scene or diorama in your classroom with the sculpture math glyphs your students create—a fantastic display for a parents' open house!

Activity Math Glyphs

Performing certain actions based on answers to a series of math questions adds some movement to math class. Activity math glyph questions should be relatively easy so that students can answer them while up and on their feet. Ask your students to move their bodies in certain ways based on their ages, heights, birthdays, or the time of day.

For example:
- raise your right hand if you are born in the first half of the year
- raise your left hand if you are born in the second half of the year
- sit down if you are 6
- stand up if you are 7
- lie down on the floor if you are neither 6 nor 7
- smile if you think 12 + 4 equals 16
- frown if you think 12 + 4 equals 18
    etc.!

Students will have fun working with numbers while moving their bodies.

Math Goofy Glyphs

Math goofy glyphs feature sensible actions for right answers, and silly actions for wrong answers. If you get the answer right, your glyph will proceed according to plan, but if you get an answer wrong, your glyph will turn out to be goofy!

Here's a math goofy glyph example in which students are creating a human figure based on glyph answers:

Which do you prefer, cats or dogs?
Answer the following math problem based on your answer:
- Cats: What is 1 + 2?
- Dogs: What is 2 + 4?

Depending on your answer, add the following to your figure:
- If your answer is 3, put a pair of running shoes on your figure.
- If your answer is 6, put a pair of boots on your figure.
- If your answer is another number, put a pair of big red clown shoes on your figure.

In this example, if your students get their answers right, their figures will end up looking as expected. But if they get their answers wrong, their glyphs will look silly, with their figures wearing, for example, big red clown shoes, a pirate hat, a curly blue wig, or a pair of pink striped pajamas.

It won't take long for most kids to realize that whenever they get the "another number" answer, they've gotten the answer wrong. This is fine, because math goofy glyphs are meant to be light-hearted. Some kids will even *want* to use the goofy answers to create their glyphs, just for fun!

Math Glyphs and the Common Core

Math glyphs give you a great opportunity to review the Common Core State Standards for mathematics for your grade level. A ten-question math glyph provides the opportunity to review several Common Core State Standards in one activity. For example, in second grade, you could include one question on addition, one question on subtraction, one question on comparing groups of objects, one question on digit place value, and so on. In this way, students will draw from a wide variety of Common Core mathematics skills to complete their math glyphs.

A math glyph lets you quickly see where your students may require extra attention based on the amount of assistance you need to give to them while they are completing their projects.

The math goofy glyphs at the end of this chapter align with these Common Core State Standards:

Kindergarten

- **CCSS.Math.Content.K.CC.A.1** Count to 100 by ones and by tens.
- **CCSS.Math.Content.K.CC.A.2** Count forward beginning from a given number within the known sequence (instead of having to begin at 1).
- **CCSS.Math.Content.K.OA.A.5** Fluently add and subtract within 5.

Grade 1

- **CCSS.Math.Content.1.OA.C.6** Add and subtract within 20, demonstrating fluency for addition and subtraction within 10. Use strategies such as counting on; making ten (e.g., 8 + 6 = 8 + 2 + 4 = 10 + 4 = 14); decomposing a number leading to a ten (e.g., 13 – 4 = 13 – 3 – 1 = 10 – 1 = 9); using the relationship between addition and subtraction (e.g., knowing that 8 + 4 = 12, one knows 12 – 8 = 4); and creating equivalent but easier or known sums (e.g., adding 6 + 7 by creating the known equivalent 6 + 6 + 1 = 12 + 1 = 13).

Grade 2

- **CCSS.Math.Content.2.OA.B.2** Fluently add and subtract within 20 using mental strategies. By end of Grade 2, know from memory all sums of two one-digit numbers.

- **CCSS.Math.Content.2.NBT.B.5** Fluently add and subtract within 100 using strategies based on place value, properties of operations, and/or the relationship between addition and subtraction.

- **CCSS.Math.Content.2.MD.C.8** Solve word problems involving dollar bills, quarters, dimes, nickels, and pennies, using $ and ¢ symbols appropriately. Example: If you have 2 dimes and 3 pennies, how many cents do you have?

Benefits of Math Glyphs

Math glyphs add to your students' learning in a number of ways:

- **Students practice math skills while doing something creative**
  Many kids enjoy working on crafts. If students believe early on that art is fun and creative, but math is boring, they may never learn to enjoy math. If you can get your early-grade students solving math problems in a fun and creative way, they'll take the excitement they feel about math with them to future grades. They'll also look for more fun ways to explore math.

- **Math glyphs result in quality finished products that are personal and unique**
  Creating glyphs gives kids a way to express themselves: glyph questions are based on their likes and dislikes, their experiences, and their personalities. Once the glyph is complete, each student will have a product to share with friends, teachers, and families. Children are very proud of their glyphs.

- **Math glyphs are an easy way to provide differentiated instruction**
  By adapting the questions on math glyphs for different students, you can change the level of difficulty to match students' skill levels. All the students in a class can work on the same glyph as their classmates while solving different math problems.

- **Math glyphs are really fun!**
  Math glyphs, especially goofy glyphs, are fun to complete. Math goofy glyphs can

have hilarious results; the more answers students get wrong, the wackier a glyph will be. Even regular glyphs can be designed to give kids silly glyph instructions. This makes completing the glyphs and comparing the finished glyphs lots of fun. Teachers who have used math glyphs in their classrooms find that students completing them often forget they're doing math!

Math Glyph Procedure

The supplies needed and the instructions to complete a math glyph are simple:

Supplies:
- Math glyph questions for each student
- For picture glyphs: crayons / colored pencils / markers
- For craft glyphs: scissors / paste or glue / colored paper / magazines
- For sculpture glyphs: modeling clay / Play-Doh
- Other art supplies as needed

Instructions:
- Photocopy one set of math glyph questions for each student before class.
- Students read each problem, solve it, then create each section of their glyph according to the directions based on their answers.

Using Completed Math Glyphs

Once glyphs have been created, there are several things to do with the results:

Discuss the glyph results in class

Part of the fun of creating a glyph is seeing the final product. And students love to look at their classmates' finished math glyphs. When all of your students have created their individual glyphs, get together as a group to view and discuss them.

You can use math glyph results to make authentic lessons in statistics. Since each glyph question has a limited number of answers, the results lend themselves well to graphing and statistical analysis. For example, if a glyph question asks students during what season their birthday takes place, you can draw a chart on the whiteboard illustrating what percentage of the class was born in each season.

Display the glyphs in your classroom

Glyphs make terrific decorations in the classroom, and students love to see their work on display, especially if that work shares a little information about themselves. Glyphs that are like "self portraits" can be used to delineate student spaces; for example, they can serve as indicators above a coat rack to show where each student will hang coats and place schoolbags, or on a classroom chore board indicating which chore each individual student will perform that day or week.

For parent conferences and open houses

Math glyphs demonstrate to parents both analytical and creative work. It's fun to let the parents guess which glyphs were created by their children. Don't forget to show the questions that were posed for each portion of the glyph (including the math involved) so that they will understand what skills their children are learning through math glyphs.

TIP

Math glyphs are fun activities for the beginning of the school year. Since the glyphs pose questions about the students and their individual interests, the glyphs students complete are snapshots of themselves at the beginning of the year. Once the glyphs are finished, students can share something interesting about themselves with their new friends.

Math Goofy Glyphs

Beginning on **page 44**, math goofy glyphs for kindergarten, grade 1, and grade 2 are included. Each math goofy glyph aligns with several Common Core State Standards for math for that grade. Print the math goofy glyph for your grade to introduce math glyphs to your students. If you prefer, you can print the pages of graphics for items that go on the puppy each student will be creating. Students can draw the items asked for in the glyphs, or they can cut out the items and paste them onto their glyphs. These puppy glyphs will serve as inspiration when you are ready to create your own math glyphs, whether goofy or regular!

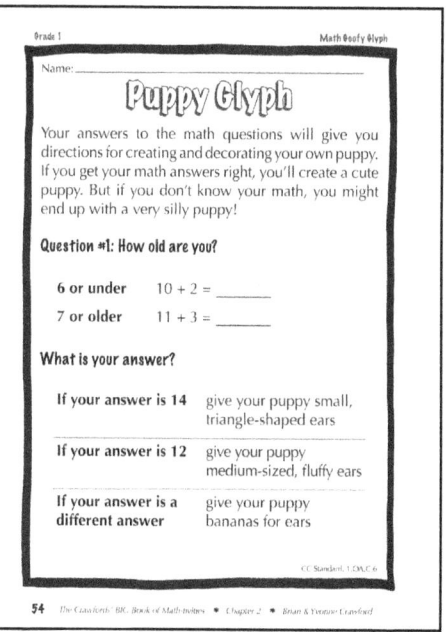

Name: _____

Puppy Glyph

Your answers to the math questions will give you directions for creating and decorating your own puppy. If you get your math answers right, you'll create a cute puppy. But if you don't know your math, you might end up with a very silly puppy!

Question #1: How old are you?

5 or under   1 + 4 = _____

6 or older   3 + 0 = _____

What is your answer?

If your answer is 5	give your puppy small, triangle-shaped ears
If your answer is 3	give your puppy medium-sized, fluffy ears
If your answer is a different answer	give your puppy bananas for ears

CC Standards: K.CC.A.1, K.CC.A.2, K.OA.A.5

Name: _____

Question #2: What is your favorite time of day?

morning 0 + 5 = _____

afternoon 1 + 1 = _____

evening 2 + 1 = _____

What is your answer?

If your answer is 3	give your puppy a small, triangle-shaped nose
If your answer is 5	give your puppy a medium-sized, circular nose
If your answer is 2	give your puppy a large, upside-down triangle-shaped nose
If your answer is a different answer	give your puppy an apple for a nose

Name: _____

Question #3: What kind of food do you like to eat?

fruits and vegetables   The number after 10 is _____

hamburgers or pizza   The number after 9 is _____

chips, nuts, or snacks   The number after 4 is _____

candy or treats   The number after 12 is _____

What is your answer?

If your answer is 11	give your puppy green eyes
If your answer is 10	give your puppy dark brown eyes
If your answer is 13	give your puppy light brown eyes
If your answer is 5	give your puppy blue eyes
If your answer is a different answer	give your puppy strawberries for eyes

Name: _____

Question #4: Do you like going to school?

yes   5 + 2 = _____

no   3 + 5 = _____

I'm not sure   4 + 5 = _____

What is your answer?

If your answer is 7	give your puppy a smiling mouth
If your answer is 8	give your puppy a frowning mouth
If your answer is 9	give your puppy a mouth with a tongue hanging out of it
If your answer is a different answer	give your puppy string beans as a mouth

Name: _____

Question #5: What games do you like to play?

sports	5 - 1 = _____
video games	6 - 3 = _____
board games	8 - 3 = _____

What is your answer?

If your answer is 3	give your puppy a short, stubby tail
If your answer is 4	give your puppy a long, thin tail
If your answer is 5	give your puppy a thick, shaggy tail
If your answer is a different answer	give your puppy a carrot for a tail

Name: _____

Question #6: What do you like doing on vacation?

play with my friends   7 - 4 = _____

visit a new place   9 - 2 = _____

go camping   6 - 2 = _____

What is your answer?

If your answer is 7	give your puppy a red collar with polka dots
If your answer is 4	give your puppy an orange collar with stars
If your answer is 3	give your puppy a blue collar with stripes
If your answer is a different answer	give your puppy a collar made of cherries

Name: _____

Question #7: What do you like to learn at school?

math and science   The number after 13 is _____

reading and writing   The number after 15 is _____

I prefer recess   The number after 16 is _____

What is your answer?

If your answer is 16	give your puppy two large spots
If your answer is 17	give your puppy four medium-sized spots
If your answer is 14	give your puppy one small spot over his eye
If your answer is a different answer	give your puppy three slices of tomato as spots

Name: _____

Question #8: What kind of toys do you play with?

action figures   8 - 1 = _____

dolls or stuffed animals   9 - 3 = _____

cars or trains   6 - 1 = _____

What is your answer?

If your answer is 7	put a bone in front of your puppy
If your answer is 6	put a stuffed dog toy in front of your puppy
If your answer is 5	put a bowl of dog food in front of your puppy
If your answer is a different answer	put a bowl of fruit in front of your puppy

Name: _____

Question #9: What is your favorite color?

red or pink   The number after 25 is _____

blue or green   The number after 27 is _____

another color   The number after 24 is _____

What is your answer?

If your answer is 26	color your puppy pink
If your answer is 25	color your puppy yellow
If your answer is 28	color your puppy light blue
If your answer is a different answer	color your puppy rainbow colors

Name: _____

Question #10: Are you a girl or a boy?

a girl $0 + 2 =$ _____

a boy $6 + 0 =$ _____

What is your answer?

If your answer is 2	title your glyph, "A Girl's Best Friend!"
If your answer is 6	title your glyph, "A Boy's Best Friend!"
If your answer is a different answer	title your glyph, "Eat Your Fruits and Vegetables!"

Finishing the Puppy Glyph

Put some finishing touches on your glyph: accessories, background, or anything else you want to add.

Name: _____

Puppy Glyph

Your answers to the math questions will give you directions for creating and decorating your own puppy. If you get your math answers right, you'll create a cute puppy. But if you don't know your math, you might end up with a very silly puppy!

Question #1: How old are you?

6 or under   10 + 2 = _____

7 or older   11 + 3 = _____

What is your answer?

If your answer is 14	give your puppy small, triangle-shaped ears
If your answer is 12	give your puppy medium-sized, fluffy ears
If your answer is a different answer	give your puppy bananas for ears

Name: _____

Question #2: What is your favorite time of day?

morning $8 + 2 + 5 =$ _____

afternoon $4 + 10 + 2 =$ _____

evening $7 + 3 + 4 =$ _____

What is your answer?

If your answer is 16	give your puppy a small, triangle-shaped nose
If your answer is 14	give your puppy a medium-sized, circular nose
If your answer is 15	give your puppy a large, upside-down triangle-shaped nose
If your answer is a different answer	give your puppy an apple for a nose

Name:_____

Question #3: What kind of food do you like to eat?

fruits and vegetables	15 - 4 = _____
hamburgers or pizza	16 - 3 = _____
chips, nuts, or snacks	12 - 6 = _____
candy or treats	14 - 9 = _____

What is your answer?

If your answer is 6	give your puppy green eyes
If your answer is 5	give your puppy dark brown eyes
If your answer is 13	give your puppy light brown eyes
If your answer is 11	give your puppy blue eyes
If your answer is a different answer	give your puppy strawberries for eyes

Name: _____

Question #4: Do you like going to school?

yes 19 - 8 = _____

no 19 - 7 = _____

I'm not sure 19 - 5 = _____

What is your answer?

If your answer is 11	give your puppy a smiling mouth
If your answer is 12	give your puppy a frowning mouth
If your answer is 14	give your puppy a mouth with a tongue hanging out of it
If your answer is a different answer	give your puppy string beans for a mouth

Name: _____

Question #5: What games do you like to play?

sports	18 - 9 = _____
video games	14 - 7 = _____
board games	16 - 8 = _____

What is your answer?

If your answer is 7	give your puppy a short, stubby tail
If your answer is 8	give your puppy a long, thin tail
If your answer is 9	give your puppy a thick, shaggy tail
If your answer is a different answer	give your puppy a carrot for a tail

Name: _____

Question #6: What do you like doing on vacation?

play with my friends   17 - 4 = _____

visit a new place   19 - 2 = _____

go camping   16 - 2 = _____

What is your answer?

If your answer is 17	give your puppy a red collar with polka dots
If your answer is 14	give your puppy an orange collar with stars
If your answer is 13	give your puppy a blue collar with stripes
If your answer is a different answer	give your puppy a collar made of cherries

Name: _____

Question #7: What do you like to learn at school?

math and science   19 + 1 = _____

reading and writing 16 + 2 = _____

I prefer recess   15 + 4 = _____

What is your answer?

If your answer is 18	give your puppy two large spots
If your answer is 19	give your puppy four medium-sized spots
If your answer is 20	give your puppy one small spot over his eye
If your answer is a different answer	give your puppy three slices of tomato as spots

Name: _____

Question #8: What kind of toys do you like to play with?

action figures	9 + 6 = _____
dolls or stuffed animals	7 + 9 = _____
cars or trains	8 + 6 = _____

What is your answer?

If your answer is 15	put a bone in front of your puppy
If your answer is 16	put a stuffed dog toy in front of your puppy
If your answer is 14	put a bowl of dog food in front of your puppy
If your answer is a different answer	put a bowl of fruit in front of your puppy

Name: _____

Question #9: What is your favorite color?

red or pink	3 + 8 + 6 = _____
blue or green	3 + 7 + 4 = _____
another color	2 + 6 + 5 = _____

What is your answer?

If your answer is 17	color your puppy pink
If your answer is 13	color your puppy yellow
If your answer is 14	color your puppy light blue
If your answer is a different answer	color your puppy rainbow colors

Name: _____

Question #10: Are you a girl or a boy?

a girl   1 + 12 = _____

a boy   16 + 1 = _____

What is your answer?

If your answer is 13   title your glyph, "A Girl's Best Friend!"

If your answer is 17   title your glyph, "A Boy's Best Friend!"

If your answer is a different answer   title your glyph, "Eat Your Fruits and Vegetables!"

Finishing the Puppy Glyph

Put some finishing touches on your glyph: accessories, background, or anything else you want to add.

Name: _____

Puppy Glyph

Your answers to the math questions will give you directions for creating and decorating your own puppy. If you get your math answers right, you'll create a cute puppy. But if you don't know your math, you might end up with a very silly puppy!

Question #1: How old are you?

7 or under   16 + 12 = _____

8 or older   14 + 12 = _____

What is your answer?

If your answer is 26	give your puppy small, triangle-shaped ears
If your answer is 28	give your puppy medium-sized, fluffy ears
If your answer is a different answer	give your puppy bananas for ears

CC Standards: 2.OA.B.2, 2.NBT.B.5, 2.MD.C.8

Name: _____

Question #2: What is your favorite time of day?

morning   32 - 16 = _____

afternoon   31 - 18 = _____

evening   39 - 13 = _____

What is your answer?

If your answer is 13	give your puppy a small, triangle-shaped nose
If your answer is 16	give your puppy a medium-sized, circular nose
If your answer is 26	give your puppy a large, upside-down triangle-shaped nose
If your answer is a different answer	give your puppy an apple for a nose

Name: _____

Question #3: What kind of food do you like to eat?

fruits and vegetables   18 + 40 = _____

hamburgers or pizza   19 + 41 = _____

chips, nuts, or snacks   15 + 42 = _____

candy or treats   17 + 42 = _____

What is your answer?

If your answer is 59	give your puppy green eyes
If your answer is 60	give your puppy dark brown eyes
If your answer is 57	give your puppy light brown eyes
If your answer is 58	give your puppy blue eyes
If your answer is a different answer	give your puppy strawberries for eyes

Name: _____

Question #4: Do you like going to school?

yes   19 + 71 = _____

no   12 + 75 = _____

I'm not sure   73 + 10 = _____

What is your answer?

If your answer is 90	give your puppy a smiling mouth
If your answer is 87	give your puppy a frowning mouth
If your answer is 83	give your puppy a mouth with a tongue hanging out of it
If your answer is a different answer	give your puppy string beans for a mouth

Name: _____

Question #5: What games do you like to play?

sports	3 nickels + 2 dimes = _____
video games	4 dimes + 7 pennies = _____
board games	6 dimes + 2 nickels = _____

What is your answer?

If your answer is 70 cents	give your puppy a short, stubby tail
If your answer is 35 cents	give your puppy a long, thin tail
If your answer is 47 cents	give your puppy a thick, shaggy tail
If your answer is a different answer	give your puppy a carrot for a tail

Name: _____

Question #6: What do you like doing on vacation?

play with friends 1 quarter + 2 dimes = _____

visit a new place 2 quarters + 4 nickels = _____

go camping 3 quarters + 12 pennies = _____

What is your answer?

If your answer is 45 cents	give your puppy a red collar with polka dots
If your answer is 87 cents	give your puppy an orange collar with stars
If your answer is 70 cents	give your puppy a blue collar with stripes
If your answer is a different answer	give your puppy a collar made of cherries

Name: _____

Question #7: What do you like to learn at school?

math and science   46 + 11 = _____

reading and writing   41 + 24 = _____

I prefer recess   45 + 14 = _____

What is your answer?

If your answer is 65	give your puppy two large spots
If your answer is 59	give your puppy four medium-sized spots
If your answer is 57	give your puppy one small spot over his eye
If your answer is a different answer	give your puppy three slices of tomato as spots

Name: _____

Question #8: What kind of toys do you like to play with?

action figures	78 - 45 = _____
dolls or stuffed animals	85 - 72 = _____
cars or trains	48 - 26 = _____

What is your answer?

If your answer is 33	put a bone in front of your puppy
If your answer is 22	put a stuffed dog toy in front of your puppy
If your answer is 13	put a bowl of dog food in front of your puppy
If your answer is a different answer	put a bowl of fruit in front of your puppy

Name: _____

Question #9: What is your favorite color?

red or pink   96 - 67 = _____

blue or green   89 - 36 = _____

another color   94 - 51 = _____

What is your answer?

If your answer is 29	color your puppy pink
If your answer is 43	color your puppy yellow
If your answer is 53	color your puppy light blue
If your answer is a different answer	color your puppy rainbow colors

Name: _____

Question #10: Are you a girl or a boy?

a girl   15 + 26 = _____

a boy   16 + 31 = _____

What is your answer?

If your answer is 41   title your glyph, "A Girl's Best Friend!"

If your answer is 47   title your glyph, "A Boy's Best Friend!"

If your answer is a different answer   title your glyph, "Eat Your Fruits and Vegetables!"

Finishing the Puppy Glyph

Put some finishing touches on your glyph: accessories, background, or anything else you want to add.

Name: _____

Puppy Glyph

Items for Puppy Glyph

Items for Puppy Glyph

Chapter 3

Word Problems + Public Speaking = Math and Tell

Math and Tell takes the fun of Show and Tell and adds math into the mix. Students tell a story to the class that illustrates a math problem. Math and Tell stories can be simple or complicated, and the math problems they illustrate can be equally straightforward or involved.

Math and Tell stories are like word problems—reversed. Instead of giving your students a word problem and having them come up with the answer, you give them the equation behind the problem and have them create the question. It's a little like the game Jeopardy!—math-style!

The goal of Math and Tell is to get students in front of the class talking about math. To create Math and Tell stories, students must understand the concepts behind the equations they are given.

A prompt for a Math and Tell story is often a full equation—for example,

$$9 - 3 = 6$$

One student might come up with the Math and Tell story: "Xavier had nine apples in his basket, gave three apples to his little sister Emily, and ended up with six apples in his basket." Not complicated! Another student might create a fantastic tale: "There were nine gigantic robots that had a battle on the surface of Mars. Three of the robots were destroyed by purple laser beams. After the battle, only six robots remained."

Both are acceptable Math and Tell stories. The goal is for students to come up with stories that show their comprehension of the underlying mathematical concepts in the prompt while being as creative as they want to be with their answers.

After they create their Math and Tell stories, the students present their stories to the class. In this way, Math and Tell offers opportunities for public speaking, while incorporating mathematical content—truly an integrated learning approach.

Creating Math and Tell Prompts

To create Math and Tell prompts, use any equation that fits with your math curriculum. Write the equation on the board if the whole class will be creating stories for the same prompt. But it's usually more fun for students to work on Math and Tell stories with a variety of prompts, so create a number of equations and give each student or each group a different one to prompt their Math and Tell presentations.

> **TIP**
>
> Math and Tell prompts can have a theme; on Valentine's Day, students can write Valentine's Day Math and Tell stories and create cards and artwork to tell them. If you're holding a class outside, your students can use rocks, leaves, and twigs as props to demonstrate their stories.

Students can also create their own Math and Tell prompts. For example, ask students to create an equation that involves subtracting two-digit numbers. They can create their own equation (for example, 90 – 58 = 32), and then come up with a story based on this equation.

Math and Tell Uses

Math and Tell can be used to reinforce new content or to review concepts as a way to check students' comprehension.

- **To reinforce content you are currently teaching**
  Hold a Math and Tell session and give several students prompts for Math and Tell stories. You will quickly see which students understand the new math concepts and which students need more instruction.

- **To review concepts previously covered**
  Create prompts for Math and Tell from equations that represent material already taught. Check the Common Core State Standards for mathematics and choose one or two equations per standard to make sure that the entire set of standards is addressed. When your students are working on their Math and Tell stories, you'll know if they have learned everything covered in those standards.

Math and Tell and the Common Core

Teach and review content from Common Core Standards for mathematics in your grade by assigning Math and Tell prompts to your students based on those Common Core Standards.

> **TIP**
>
> Create a Math and Tell book or wall display containing the various Math and Tell stories that your students have presented throughout the year. Align those Math and Tell stories with the Common Core Standards that your students are studying during the year. By the end of the year you will have a helpful display that can aid in reviewing and reinforcing important mathematical skills.

To ensure that all the Standards for your grade are covered, keep a checklist of Common Core Standards for your grade and write down the dates that they were presented to the class via a Math and Tell presentation, including the names of the students who gave the presentations.

Benefits of Math and Tell

- **Presentations are fun and creative**
  Bringing an element of fun to math class is important. If you feature a Math and Tell session every now and then at the beginning of class, it will kick off your math class in an upbeat way.

- **Show true understanding of the mathematical concepts**
  In order to answer math problems, sometimes students can rely on memorization or standard algorithms. But to create and present Math and Tell stories, students must understand why equations operate the way they do. When your students create and present the stories behind the math problems they are given, you will help them develop number sense and will know if they understand the concepts behind the math problems.

- **Gives students an opportunity to present in front of the class**
  Presenting Math and Tell stories allows students to practice their presentation, acting, reasoning, and storytelling skills. It's an opportunity to integrate your instruction.

- **Tailored to fit individual students and their abilities**
  Math and Tell can be used by kids of all ages and skill levels. Younger students can even draw their Math and Tell stories if they are not yet able to write them out. Older students can work in groups cooperatively to act out the stories they create.

  Math and Tell is a great strategy for differentiated instruction. It can be tailored to fit the individual student who may need help practicing basic skills or one whose math skills are more advanced.

Presenting Math and Tell Stories

After students create their Math and Tell stories, they can present them to the class in several different ways:

- **As a speech**
  Students can stand in front of the class and tell their stories.

- **As a play**
  Students working in groups can act out their stories.

- **Using props**
  Students can create props, compose drawings, build sculptures, or gather manipulatives to demonstrate their stories to the class. Let your students' creativity flow (within the bounds of time, cost, and safety, of course) while they create their "math"terpieces!

Doing a presentation in front of the class can be nerve-wracking! When students finish their Math and Tell presentations, encourage the audience to applaud.

TIP

Review each student's Math and Tell story before they give their presentations. If a student is struggling during the presentation, jump in and help out! Do what you can to help your students succeed with their presentations. Remember, presenting in front of a class is another new skill for them to learn.

Chapter 4

Math Games and Puzzles

Games and puzzles bring challenge, competition, and cooperation to math class. Students respond well to math presented as a game to play or a puzzle to solve.

Challenge:
Math games and puzzles present fun problems for students that work well to practice skills and learn new ones.

Competition:
Multi-player games spark competition between students, which can bring excitement to your lessons. Playing competitive games with math will often give children a reason to try to excel at mathematics and to improve on their math skills.

Cooperation:
Some math games are cooperative, not competitive. In these games, students work together toward a common goal and build a spirit of teamwork while working on math problems.

When playing games in the classroom, the focus should be on the fun and challenge of the games, and not on who is winning the games or whose math skills are the strongest. This is particularly important to keep in mind if your students enjoy competitive games.

Benefits of Math Games and Puzzles

- **Adding fun to learning**
Math games are exciting activities that help teach students valuable skills and practice what they have learned. The most effective math games get kids excited to learn, not frustrated or put off by the math inherent in the game or puzzle.

- **Skills practice**
Students need to practice new math skills, and math games are very effective for making math practice fun. For most students, a game or puzzle is far more appealing than worksheets or other drill methods.

* **Flexible teaching tools**

Math games can be used for a variety of educational purposes. Different math games can be used to practice a multitude of skills, and games can be created to challenge different students in different ways.

Even a very simple math game, such as a game that has students complete a color-by-numbers activity, can align with several Common Core Standards. For kindergarten students, the boxes in the color-by-numbers puzzle might simply contain numbers. For a first grade student, the boxes in that same puzzle might feature simple addition equations. For second grade students, slightly more complicated addition or subtraction problems could be used.

Math puzzles are very useful when providing differentiated instruction. If the students in your classroom have different levels of mathematical ability, or if you're teaching a multi-grade classroom, adjust a math puzzle for three levels of ability: basic, intermediate, and advanced. Give each student the version of the math puzzle that they will find challenging and enjoyable. Students can work on their version of the math puzzle without knowing that they're completing something different from the other kids in the class.

* **Homework**

Math puzzles and games make great homework assignments that your students will enjoy completing.

When and Where to Use Math Puzzles and Games

* **At the beginning of the school year**

Math games are a great fit for the start of the school year. You can use games as icebreakers for the class or as ways to introduce new students to one another. Getting students moving and talking with each other helps to build a cohesive class.

You can also use math games early in the school year as a way to assess your students' skills, determining to what extent they have mastered the skills from the Common Core Standards from their previous year of school, and to what extent they know this year's skill set.

Any entertaining game or puzzle makes for a good activity for the beginning of the school year. The goal is to make early math lessons fun and interactive to set the tone for the rest of the year.

* **Before holidays or vacations**

When holidays like Halloween or Christmas are just around the corner, kids can have a hard time concentrating on schoolwork. At times like these, playing games can help kids release their pent-up energy. A fun holiday-themed math game to play in your classroom or outside in the schoolyard will help them continue to learn despite the distraction of the upcoming holiday.

* **When the kids need something different**

You'll know when your students need a break from their daily routine. When your students reach a point where they're no longer learning effectively, this may be a good time to introduce a new math game.

- **As a unit review**
After you've taught a unit, math games are a great way to review the skills learned. Bring back games over time to continue to review skills learned earlier.

- **As an activity for early finishers**
Early finishers shouldn't feel like they're being punished for being faster than the other students in their class by getting even more work to complete. But when you provide them a fun and challenging math puzzle, it will feel like a reward, not a punishment.

- **To demonstrate to others what your students can accomplish**
You can use games when you are being audited or assessed. Games are a good way to show your students working and interacting, and they help to demonstrate the energy and dynamic nature of your class.

- **At the end of the school year**
At the end of the year, games can be a fun way to "bring it all together." Use a variety of different games that incorporate the math skills that your students have acquired during the school year, or a larger game that involves multiple math skills.

Math Games

The math games in this chapter can be used as solo games that students complete at work centers or games that a group of students works on cooperatively. You can also use these games in a competitive manner—for example, have students try to be the first to complete a challenge or solve a puzzle.

1 Math Libs

Math Libs are games where players choose words—nouns, verbs, pronouns, and adjectives—to fill in the blanks of a fun story. Math Libs are like "Mad Libs™," but the student must answer math problems to determine what word fills in the blank.

For example, one part of a Math Libs story might be:
On her visit to the farm, Isabella saw a field full of _____.
   1 + 3 =
   3: Apple trees
   4: Cows
   6: Porcupines
   8: Spaceships

The student solves the math problem in order to find the correct word or phrase to write on the blank line. Once all of the math problems have been solved and answers revealed, the student can read the story from start to finish.

Math Libs are entertaining, potentially silly activities. They have the added advantage of giving students practice in both mathematics and English language arts skills. You and your students can create Math Libs stories about your own school or town, or even about the individual students in your class.

Beginning on **page 89**, there are Math Libs stories for kindergarten, grade 1, and grade 2 that each align with a specific Common Core Standard, plus a blank template (**page 92**) to create your own Math Libs stories.

2 Picture Math

Picture Math puzzles are color-by-number puzzles featuring a variety of different math problems. The individual shapes to be colored feature math problems to be solved to determine the color for each shape.

For example, a key might feature the following number-to-color translations:

   2: Red
   4: Blue
   5: Green
   8: Brown

On the puzzle, each individual shape would have a different math problem, such as:

   1 + 1 =
   4 + 1 =
   3 + 2 =
   3 + 5 =

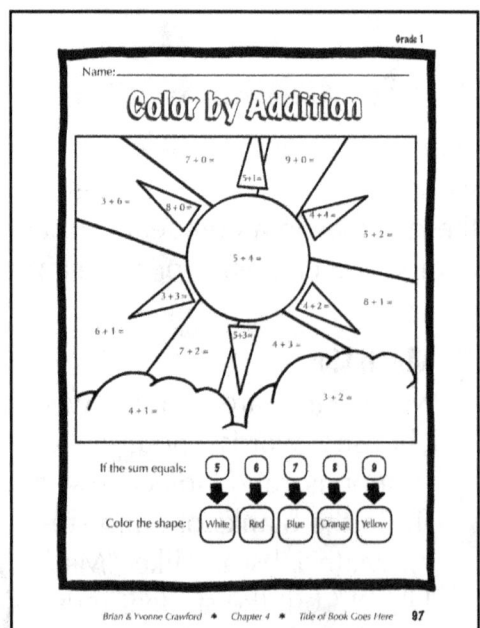

Students solve the problems in the shapes to determine which color they need to color in each shape according to the color key.

You can populate the shapes in a picture with a variety of different math problems depending on what math skills you are learning or reviewing with your students.

For example:
   Kindergarten: Simple numbers (1, 2, 3, 4…)
   First grade: Simple addition and subtraction problems (4 – 2 =____)
   Second grade: More complex addition and subtraction problems
   (24 + 12 + 3 =____)

Beginning on **page 93**, there are Picture Math puzzles for kindergarten, grade 1, and grade 2 that each align with a specific Common Core Standard, plus a blank template (**page 99**) for your own Picture Math puzzle creations.

3 Bubble Math

Bubble Math puzzles are simple to create and fun to complete. Students are presented with a page full of bubbles (or squares, or some other shape) with a variety of different numbers on them. Students compare a given number with the number in each bubble. They fill in the bubbles or leave them blank according to a stated rule.

For example, a rule might be to color in each bubble with a value of less than or equal to 10. Each bubble would contain a number. According to the rule, the bubble with a 6 would be colored in by the student, but the bubble with a 12 would not. Once all of the bubbles are colored in, the student would then see the resulting picture "hidden" in the field of bubbles.

Bubble Math puzzles can be created using a template to correspond with holidays or special events at your school. They are also fun to play competitively; students race to be the first to find the picture hidden in a puzzle.

Beginning on **page 96**, there are Bubble Math puzzles for kindergarten, grade 1, and grade 2 that each align with a specific Common Core Standard, plus a blank template (**page 99**) for a Bubble Math puzzle that you can create.

4 Math Find and Draw

Math Find and Draw puzzles strengthen students' skills in arithmetic and spatial awareness while adding the fun of drawing or cutting and pasting simple pictures.

The first page of a Math Find and Draw puzzle is a series of math problems. The second page has numbers printed in different places on the page. Students answer each math problem on the first page. When they find the solution to a problem, they are directed to draw a picture of an item at that number on the second page. An alternate method is to have students cut out and paste the items asked for on the first page (the items needed for each puzzle are included after the puzzle). There are more numbers on the second page than are needed for the answers, so students will need to look carefully to find the correct numbers. When they have completed the whole puzzle, they will have a full-page scene.

Beginning on **page 100**, there are Math Find and Draw puzzles for kindergarten, grade 1, and grade 2 that each align with a specific Common Core Standard.

5 Math Number Search

Math Number Search puzzles are like word search puzzles, but players search for numbers instead of words.

Students answer math problems. Once they find an answer to each question, they search for that answer in a number grid.

Math number search puzzles can be created for different grade levels. They are easy to correct by comparing the completed grids to an answer sheet grid for a quick visual confirmation of correctly solved problems. When you create math number search puzzles, be sure to account for the possibility of multiple correct answers to be found within a grid.

Beginning on **page 109**, there are Math Number Search puzzles for kindergarten, grade 1, and grade 2 that each align with a specific Common Core Standard.

6 Match the Math

Match the Math is a matching game that lets students build a complete equation with game cards.

Students choose one answer card, then select two parts of the equation to make the answer true. For example, in the kindergarten version of the game (**pages 112-114**), the student selects one worm answer card, then finds two apple halves that go together to create an equation that equals the number on the worm card.

Teacher set-up for these games is easy: print, laminate, and cut out the cards for each game. The student stacks all the answer cards, and spreads out the "half" cards on the desk. To play, the student selects one answer card, then searches the half cards to find the two pieces that together make an equation equal to the answer card.

Match the Math can be played individually, in a center, or together as a class. Beginning on **page 112**, there are Match the Math games for kindergarten and grade 1 that each align with a specific Common Core Standard. The game for grade 2 is left blank (pages 118-119) so teachers can create their own cards to meet a variety of Common Core Standards for mathematics.

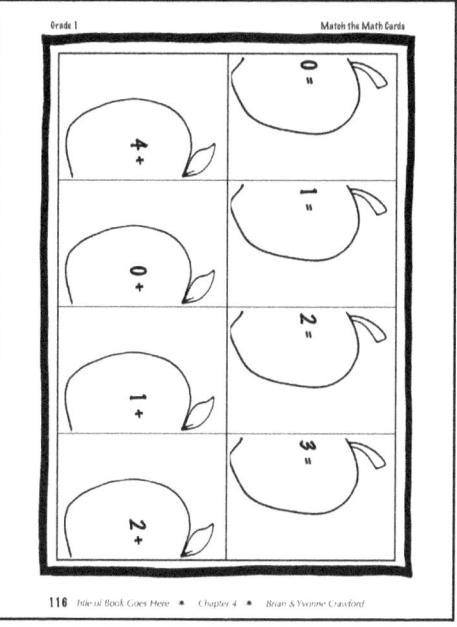

Creating Your Own Math Games

You can create your own math games to help reinforce the subjects you are teaching in class. By providing a variety of different types of games and puzzles to your students, you will be offering many entertaining challenges for them to complete.

Several templates are included in the math games described in this chapter to create your own games for your math curriculum. If you enjoy the creative process, you can also use a document publishing program such as Publisher, InDesign, or QuarkExpress, or a presentation graphics program such as PowerPoint to create your own games and puzzles.

Another way to create a math game is to take an existing game and work math problems into it. Here are a few examples:

1 Math Go Fish

Take out the jacks, queens, kings, and jokers from a deck of cards (Ace = 1). Instead of asking, "Do you have any eights?" a player would ask, "Do you have any five plus threes?" The other player would have to solve the math problem in his or her head, come up with the answer of eight, and then decide whether they should give the other player a card or say "Go fish!"

2 Math Hopscotch

Draw a hopscotch board on the ground and add numbers to the board. As students jump from number to number, they can either count each number in sequence to strengthen their counting skills, or add the number they hop on to their previous total to practice their addition skills. For example, a student might begin a game of hopscotch by hopping on the number 1, then hop to the number 2. Then the student would pause and say, "one plus two equals three!" As an added challenge, have students *subtract* numbers from their total as they work their way backwards.

3 Math Memory

Create a set of Math Memory cards using blank 3 x 5" file cards. The cards that "match" are as follows: one card with the answer (for example, **4**), and the matching card with the math problem (for example, **3 + 1**). Make 10 to 20 matching pairs for a full deck. To play the game, all the cards are laid face down in a grid on a table or desk. Player 1 flips up two cards, looking for a matching pair of cards (one that shows the equation and the other that shows the answer). If Player 1 finds a match, he/she takes that pair of cards off the table and gets another turn. If Player 1 doesn't find a match, he/she returns the cards to the same spot and Player 2 takes a turn.

Name:_____

Math Libs

The Missing Items

Mason was looking for his _____. He looked under the
                                    A

_____ and behind the door. He couldn't find them anywhere.
   B

All of a sudden, he looked _____ and saw them! They
                              C

were right where they should be—on his feet!

A 5 - 1 = _____	B 7 - 6 = _____	C 5 - 0 = _____
3 - pet rocks	1 - bed	4 - up
4 - socks	2 - dog	6 - all around
5 - twin sisters	3 - kitchen sink	5 - down

CC Standard: K.OA.A.1

Name: _____

Math Libs

The Missing Items

Mason was looking for his _____. He looked under the
                                    A

_____ and behind the door. He couldn't find them anywhere.
    B

All of a sudden, he looked _____ and saw them! They
                               C

were right where they should be—on his feet!

A 9 - 6 = _____	B 17 - 6 = _____	C 15 + 5 = _____
4 - pet rocks	11 - bed	19 - up
3 - socks	12 - dog	21 - all around
5 - twin sisters	13 - kitchen sink	20 - down

CC Standard: 1.OA.C.6

Math Libs

The Missing Items

Mason was looking for his _____. He looked under the
     A

_____ and behind the door. He couldn't find them anywhere.
   B

All of a sudden, he looked _____ and saw them! They
                              C

were right where they should be—on his feet!

A 29 - 16 = _____	B 37 - 16 = _____	C 35 + 15 = _____
14 - pet rocks	21 - bed	49 - up
13 - socks	22 - dog	51 - all around
15 - twin sisters	23 - kitchen sink	50 - down

CC Standard: 2.NBT.B.5

Name:_____

Math Libs

1	2	3

Kindergarten

Name:_____

Color by Numbers

If the number equals: 5 6 7 8 9 10

Color the shape: Green Red Blue Orange Yellow Pink

CC Standard: K.CC.A.1

Grade 1

Name:_____

Color by Addition

7 + 0 =

9 + 0 =

5 + 1 =

3 + 6 =

8 + 0 =

4 + 4 =

5 + 2 =

5 + 4 =

3 + 3 =

4 + 2 =

8 + 1 =

6 + 1 =

5 + 3 =

7 + 2 =

4 + 3 =

3 + 2 =

4 + 1 =

If the sum equals: 5 6 7 8 9

Color the shape: White Red Blue Orange Yellow

CC Standard: 1.OA.C.6

94   The Crawfords' BIG Book of Math-tivities   ✱ **Chapter 4** ✱   Brian & Yvonne Crawford

Name: _____

Color by Addition

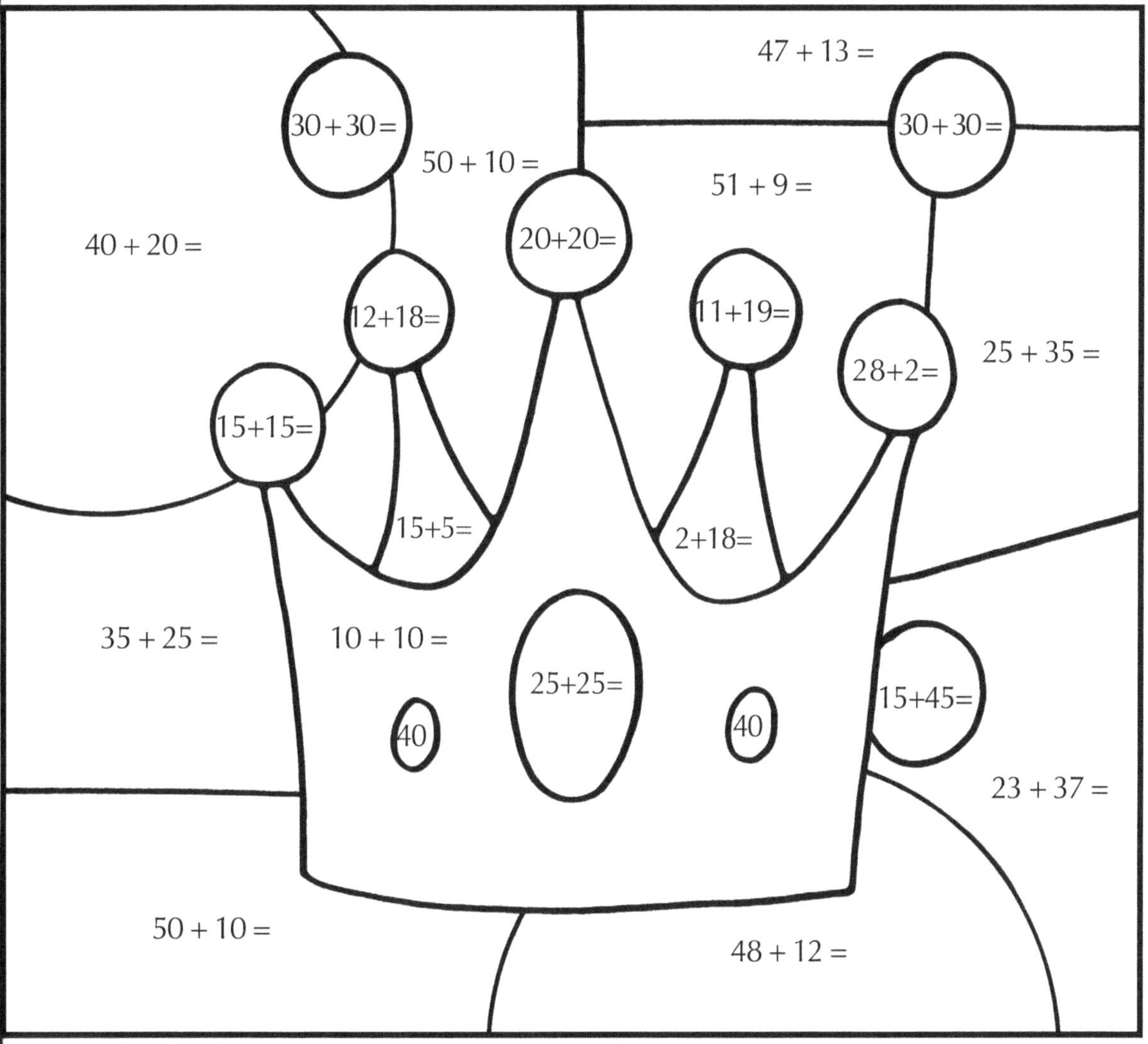

$30 + 30 =$

$47 + 13 =$

$50 + 10 =$

$30 + 30 =$

$40 + 20 =$

$51 + 9 =$

$20 + 20 =$

$12 + 18 =$

$11 + 19 =$

$25 + 35 =$

$28 + 2 =$

$15 + 15 =$

$15 + 5 =$

$2 + 18 =$

$35 + 25 =$

$10 + 10 =$

$25 + 25 =$

40

40

$15 + 45 =$

$23 + 37 =$

$50 + 10 =$

$48 + 12 =$

If the sum equals:

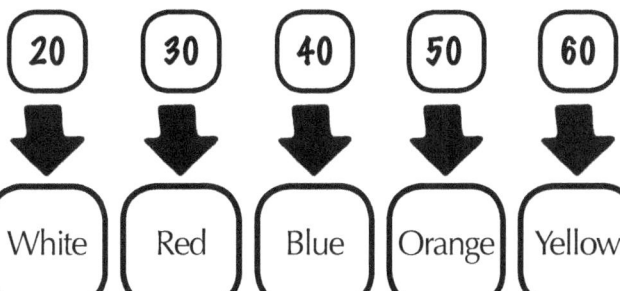

CC Standard: 2.NBT.B.5

Grade 2

Kindergarten

Name:_____

Bubble Math

Color the number bubbles that are greater than or equal to 10

3	9	1	0	5	2
5	4	17	15	8	9
8	14	5	2	11	8
4	13	3	0	19	3
0	12	4	5	20	2
8	1	16	18	8	0
2	3	0	9	8	0

CC Standard: K.CC.C.7

Bubble Math

Color the number bubbles that are greater than or equal to 30

23	19	31	5	15	22
15	20	43	52	18	29
18	28	55	42	43	28
14	16	35	20	10	13
40	46	42	35	32	22
8	39	32	35	18	20
12	23	20	29	18	10

CC Standard: 1.NBT.B.3

Grade 2

Name:_____

Bubble Math

Color the number bubbles that are greater than or equal to 100

38	38	30	38	149	69	74	28	61
42	23	42	59	177	179	26	64	57
39	35	26	83	151	132	142	37	43
88	64	24	18	111	120	119	142	60
37	79	68	99	110	16	17	55	41
66	24	30	71	146	48	11	50	66
40	46	17	22	163	26	52	71	58
39	136	114	119	122	137	220	152	65
80	19	102	106	110	118	117	7	58
61	53	49	89	15	29	65	32	45

CC Standard: 2.NBT.A.4

Name: _____

Bubble Math

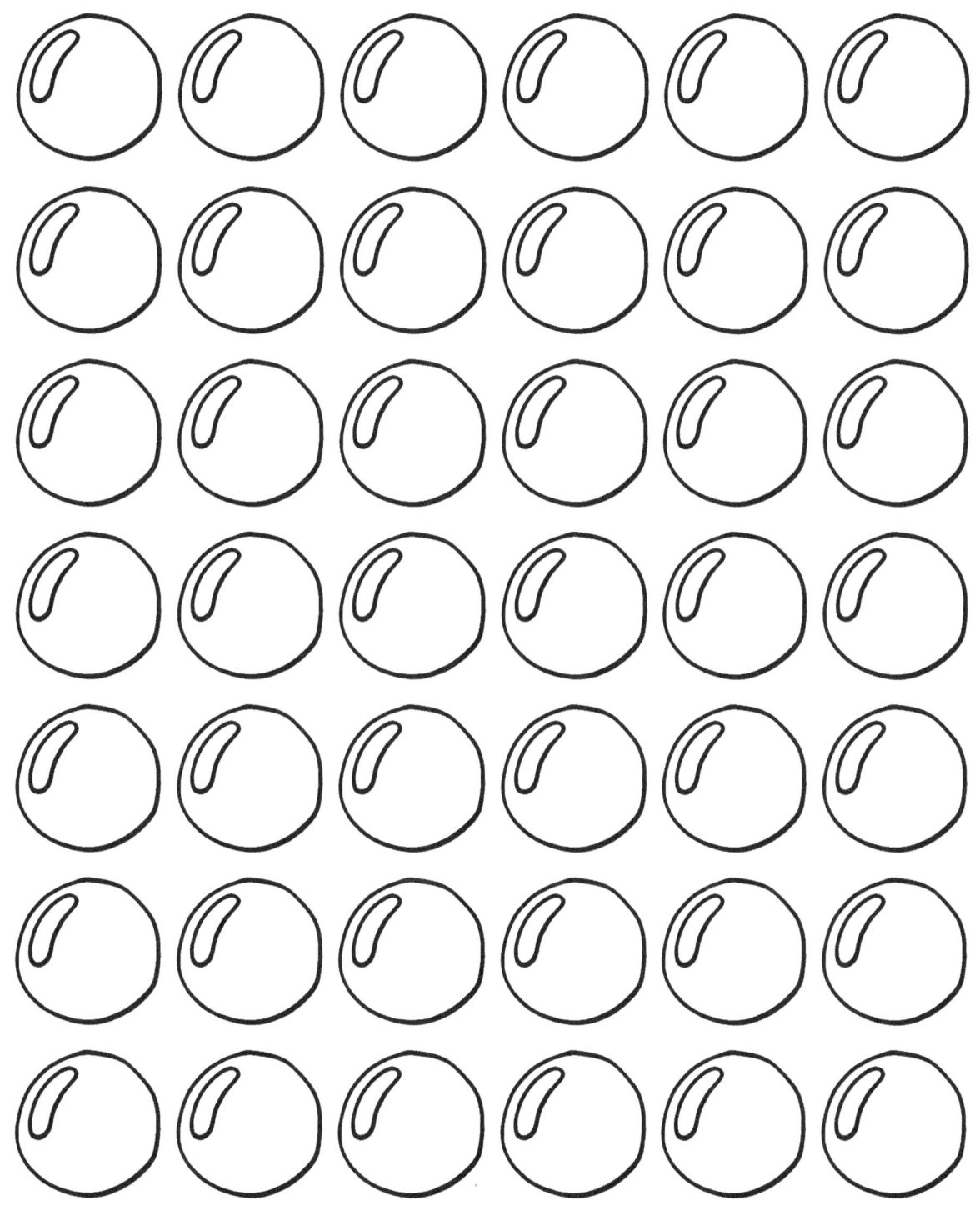

Kindergarten

Name: _____

Garden Find & Draw Puzzle

Answer each problem. Then find your answer on the puzzle page, and draw the garden items near the number of your answer.

1 + 1 = Find this number and draw a flower	**5 + 0 =** Find this number and draw a tree
0 + 3 = Find this number and draw 2 carrots	**1 + 5 =** Find this number and draw a watering can
0 + 0 = Find this number and draw a sun	**0 + 1 =** Find this number and draw a cat
1 + 3 = Find this number and draw 1 child	**3 + 4 =** Find this number and draw a dog

CC Standard: K.OA.A.5

Name: _____

Garden Find & Draw Puzzle

Kindergarten

0

8

9

5   4

3

1   2

6   7

Brian & Yvonne Crawford * **Chapter 4** * The Crawfords' BIG Book of Math-tivities

Kindergarten

Grade 1

Name: _____

Pet Find & Draw Puzzle

Answer each problem. Then find your answer on the puzzle page, and draw the pets near the number of your answer.

5 + 1 = Find this number and draw a cat	**9 + 2 =** Find this number and draw a bunny
2 + 3 = Find this number and draw a dog	**9 + 3 =** Find this number and draw a sun
9 + 1 = Find this number and draw a cloud	**2 + 6 =** Find this number and draw a bird
4 + 3 = Find this number and draw a flower	**5 + 4 =** Find this number and draw a turtle

CC Standard: 1.OA.C.6

Brian & Yvonne Crawford * Chapter 4 * The Crawfords' BIG Book of Math-tivities

Grade 1

Name: _____

Pet Find & Draw Puzzle

10

12

8

15

11

7

9

5

6

18

Grade 1

Grade 2

Name: _____

School Find & Draw Puzzle

Answer each problem. Then find your answer on the puzzle page, and draw the school items near the number of your answer.

15 + 21 = Find this number and draw a chalkboard	**9 + 12 =** Find this number and draw 2 books
12 + 3 = Find this number and draw a student	**19 + 3 =** Find this number and draw a window
19 + 1 = Find this number and draw a teacher	**5 + 12 =** Find this number and draw 3 rulers
14 + 13 = Find this number and draw a pencil	**15 + 14 =** Find this number and draw a desk

CC Standard: 2.NBT.B.5

Name: _____

School Find & Draw Puzzle

Grade 2

18   36

                                    22

        20

                                    27
                        29

    15

                17

                            21

10

Brian & Yvonne Crawford * Chapter 4 * The Crawfords' BIG Book of Math-tivities

Grade 2

108 The Crawfords' BIG Book of Math-tivities * **Chapter 4** * Brian & Yvonne Crawford

Name:_____

Number Search

Solve the problems below, then find the answers to the problems in the number search grid.

1. 3 +1 ___
2. 0 +1 ___
3. 0 +0 ___
4. 0 +2 ___
5. 1 +4 ___

6. 1 +0 ___
7. 0 +0 ___
8. 1 +1 ___
9. 2 +1 ___
10. 1 +1 ___

1	0	8	2
9	4	7	2
7	3	1	6
0	2	5	5

CC Standard: K.OA.A.5

Name: _____

Number Search

Solve the problems below, then find the answers to the problems in the number search grid.

1. 8
  +1
  ___

2. 5
  +1
  ___

3. 0
  +2
  ___

4. 6
  +2
  ___

5. 2
  +5
  ___

6. 4
  +1
  ___

7. 1
  +3
  ___

8. 0
  +1
  ___

9. 2
  +1
  ___

10. 1
  +1
  ___

0	5	8	9
6	0	3	0
0	2	0	2
4	0	1	7

CC Standard: 1.OA.C.6

Name: _____

Number Search

Solve the problems below, then find the answers to the problems in the number search grid.

1. 10
  +23
  ___

2. 26
  +13
  ___

3. 25
  +12
  ___

4. 16
  +22
  ___

5. 21
  +35
  ___

6. 23
  +11
  ___

7. 19
  +10
  ___

8. 24
  +11
  ___

9. 51
  +10
  ___

10. 30
  +11
  ___

2	9	3	0	3
0	5	7	3	6
0	3	0	5	1
3	4	4	6	9
0	8	1	3	0

CC Standard: 2.NBT.B.5

Kindergarten

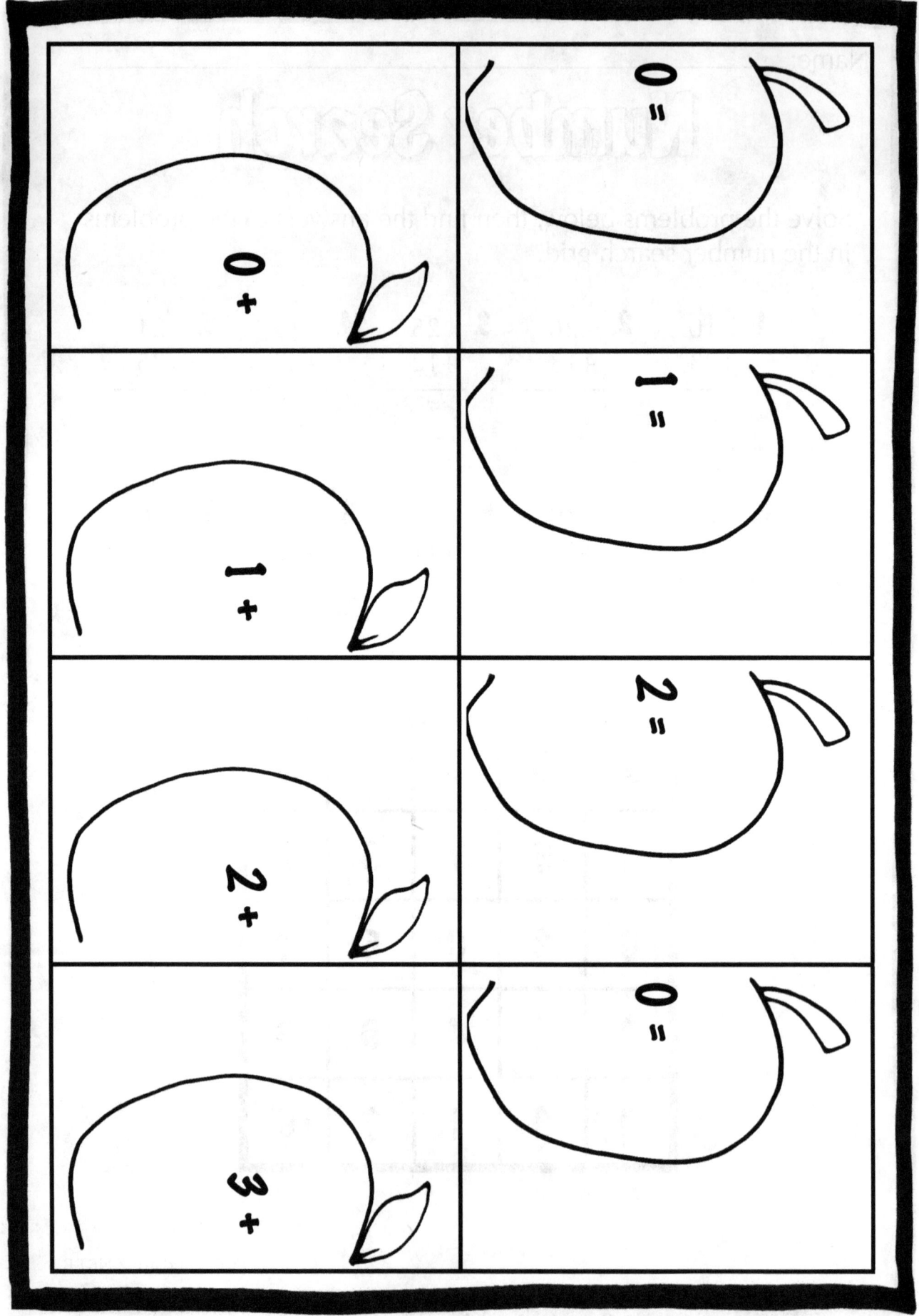

112 The Crawfords' BIG Book of Math-tivities   * **Chapter 4**   * Brian & Yvonne Crawford

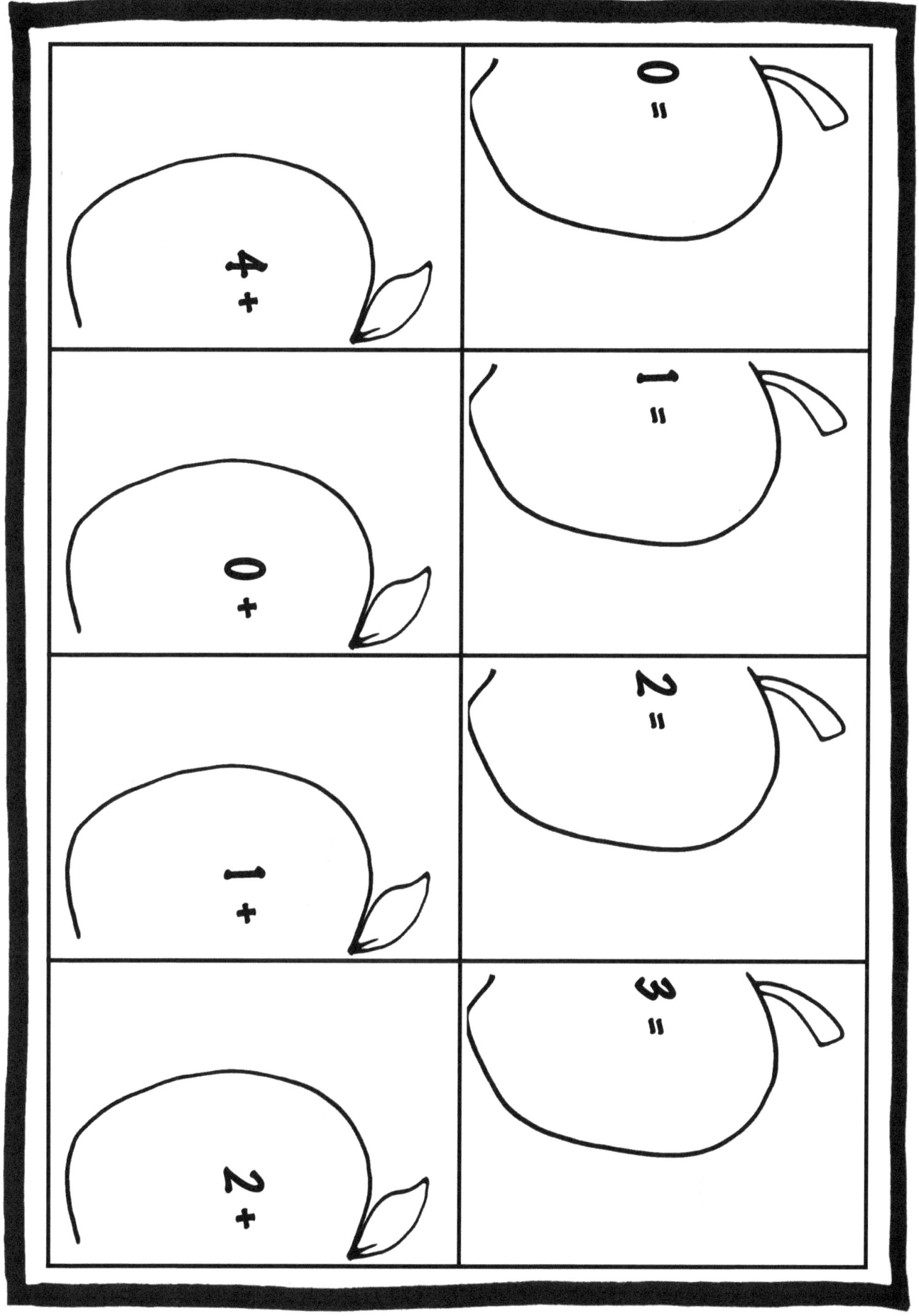

Kindergarten

CC Standard: K.OA.A.5

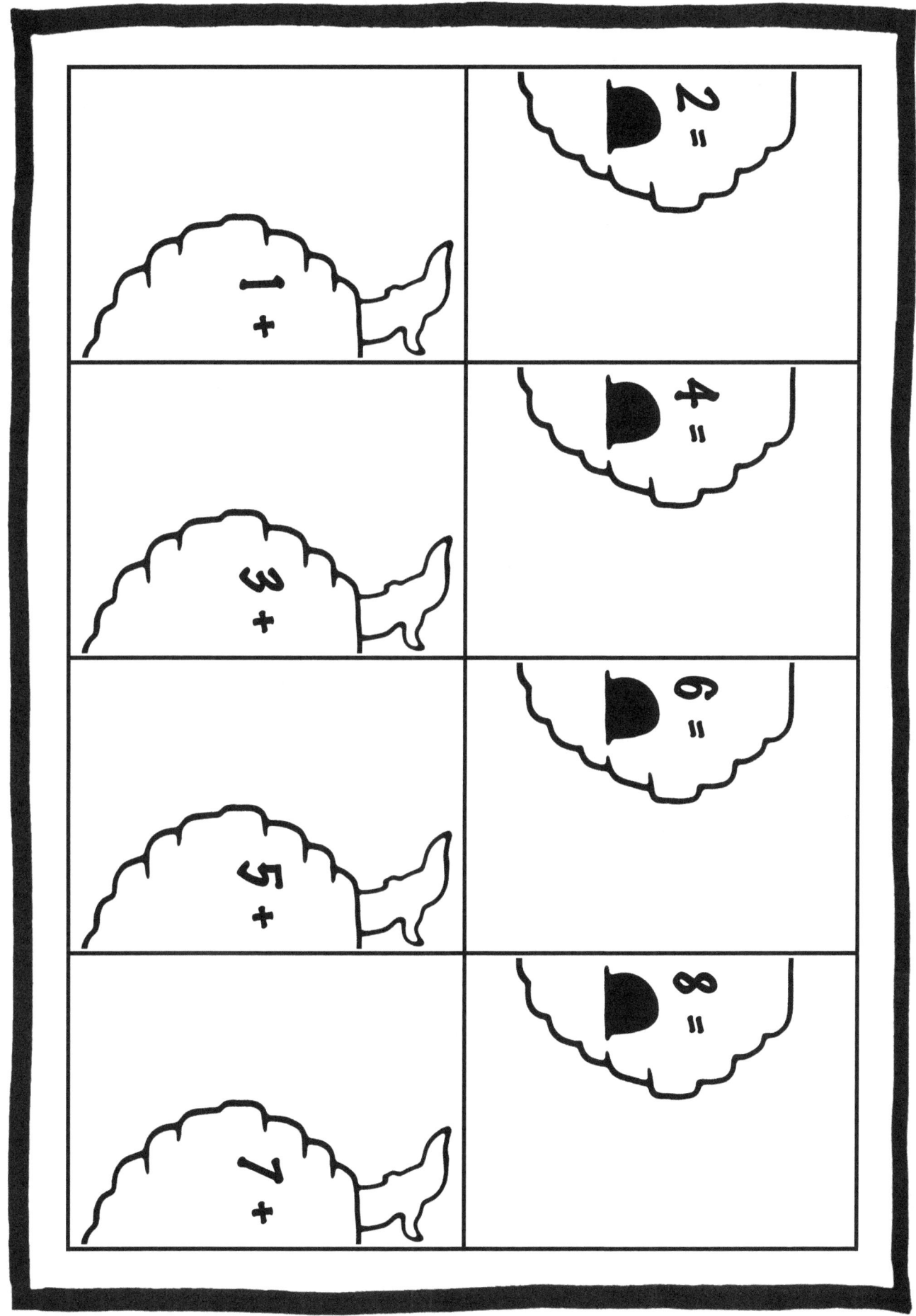

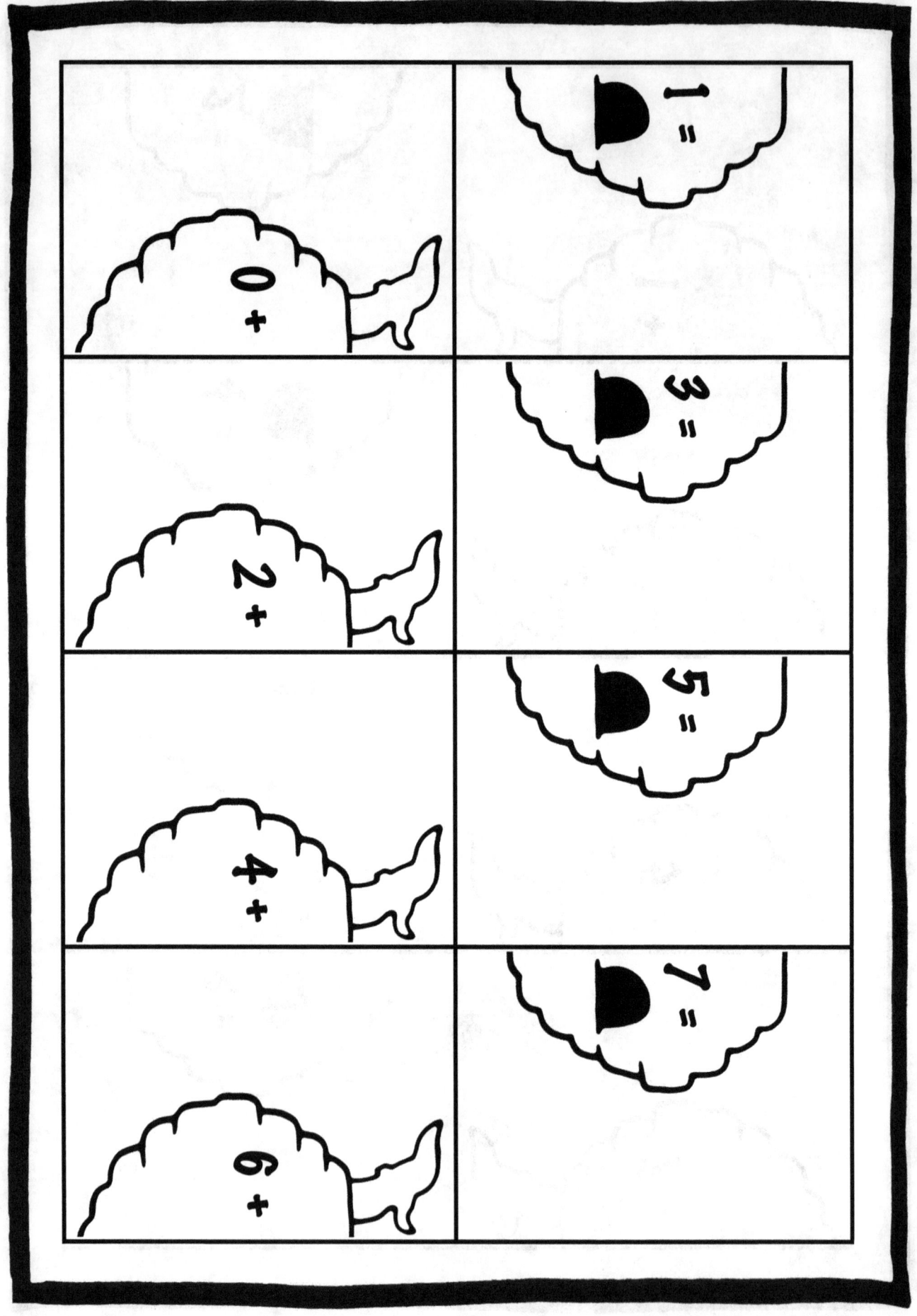

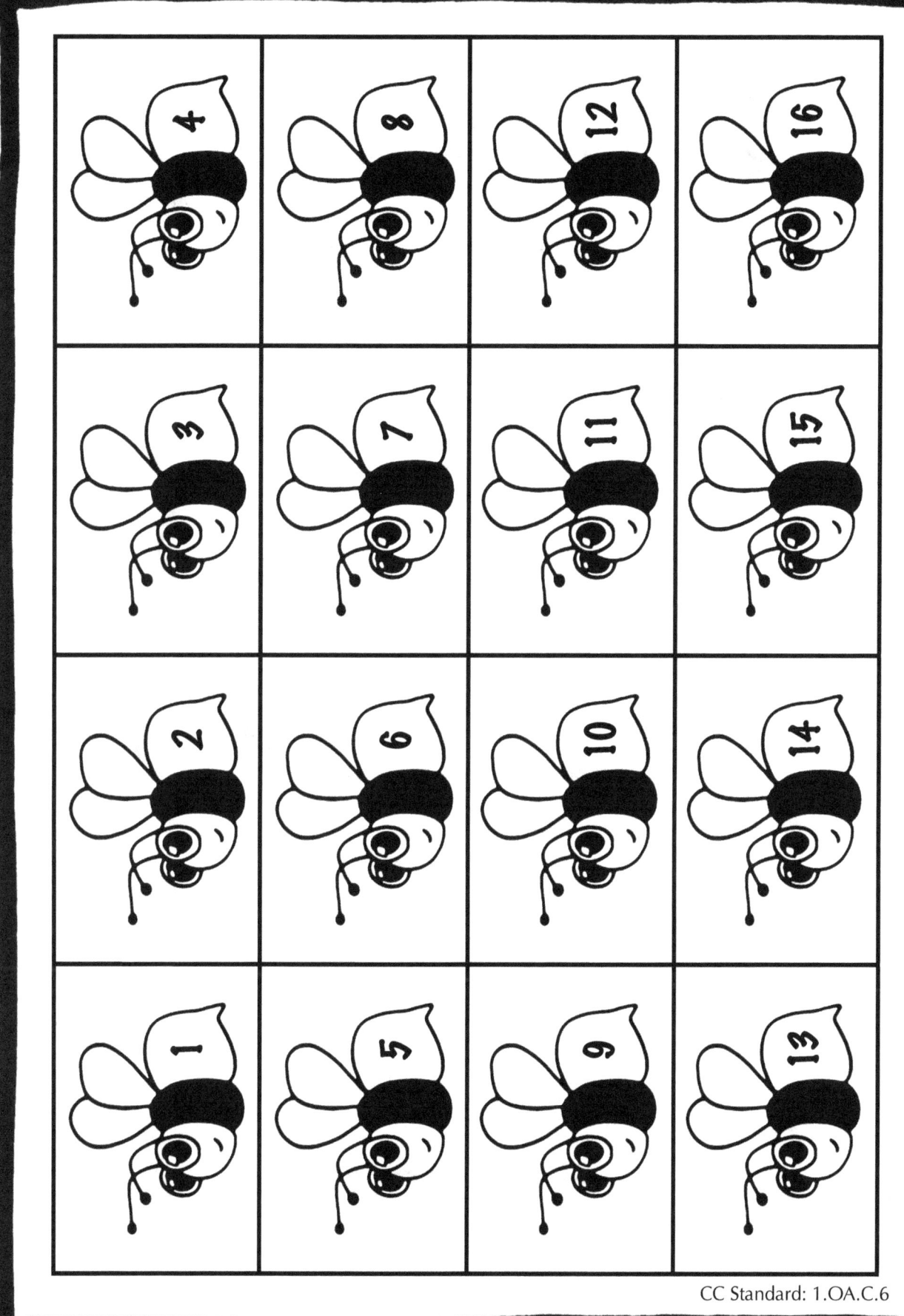

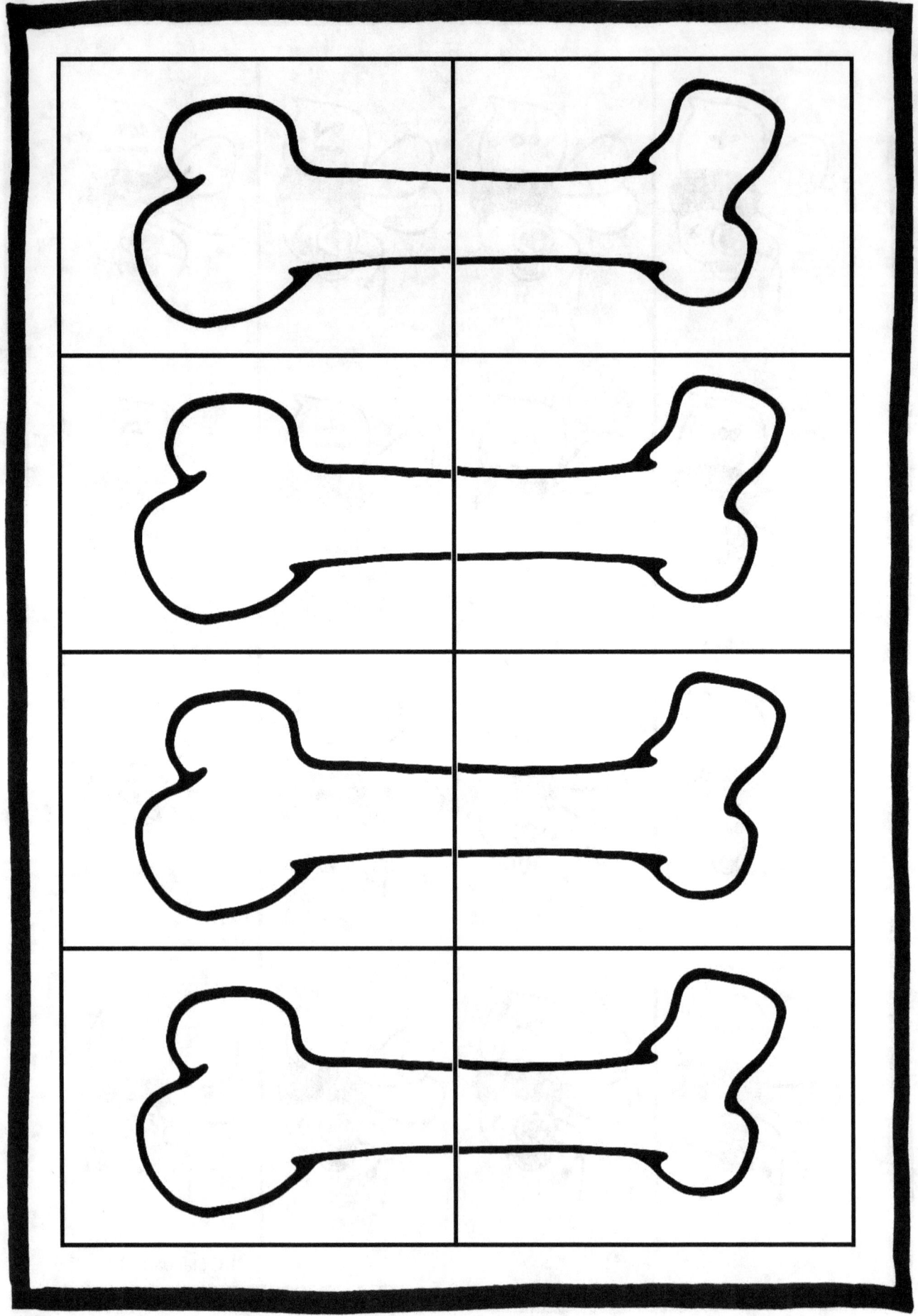

Chapter 5

Taking Math Outside

Bringing your students into a new environment to learn can refresh their minds and open them to new ways of learning. Beyond the general delight in being outdoors, when you hold math class outside, you can show your students where mathematics can be found in nature.

Benefits of Taking Math Outside

- Exercise kids' bodies and refresh their minds while learning math
- Incorporate outdoor science into your math lessons
- Give children (and teachers) a welcome change of scenery
- Provide opportunities to bring multiple classes together and for teachers to work with other teachers

TIP

Before teaching a class outside, take a walk around your school or playground to note the opportunities available to teach mathematics.

Activities

1 Mathbook or math journal prompts for outdoors

If you use math journals or Mathbooks, create prompts that must be answered outside. These prompts may include questions such as:

- How many trees can you find in the playground?
- How many steps does it take to cross the baseball diamond?
- Look at your watch. What time is it? Now run across the soccer field. What time is it now? How many minutes did it take for you to run from one end of the soccer field to the other?

- Find a fall leaf and paste it in your Mathbook. Now measure the leaf. How tall is it? How wide is it?
- Look at the school. Subtract the number of doors you see from the number of windows.
- How many cars are in the parking lot? Without counting the number of tires on each car, can you figure out how many tires there are in total?

Beginning on **page 126**, there are Mathbook or math journal prompts for outdoor activities for kindergarten, grade 1, and grade 2 that align with Common Core Standards.

2 Hide, Seek, and Solve

Hide, Seek, and Solve is a math game that can be played either indoors or outside, but taking the game outside is especially fun.

Here's how to play:

- Print, laminate, and cut out the Hide, Seek, and Solve cards, using as many cards as you'd like for your class.
- Glue Popsicle sticks to the cards and/or punch holes in them and thread yarn or string through the holes to hang them.
- Place the cards in various spots outside or around your classroom. Hide the cards:
    - Outside—hang cards from tree branches, stick them into the outside of flower beds, wedge them between rocks, etc.
    - Inside—hide cards in cabinets, tie them to chairs or desks, hang them on door knobs or cabinet handles, staple them to bulletin boards, etc.

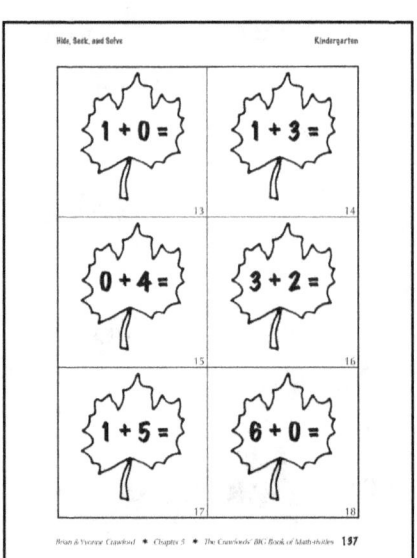

- Give each student a tally sheet (**page 132**) to record the answers. Clipboards are helpful if you have them.
- Students move around the area where the cards are hidden, looking for Hide, Seek, and Solve cards. When they find a card, they solve the math problem and record their answer on the tally sheet.

- The game is over when (your choice):
    - the first student solves all the questions correctly OR
    - all the students have answered all the questions OR
    - time is up

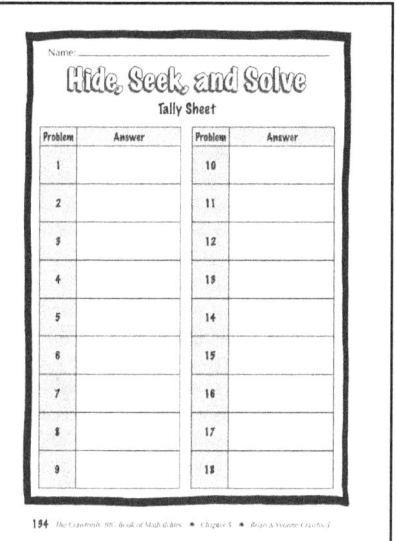

Beginning on **page 133**, there are sets of Hide, Seek, and Solve cards for kindergarten, grade 1, and grade 2. Each set aligns with the following Common Core Standards:

- Kindergarten - K.OA.A.5
- Grade 1 - 1.OA.C.6
- Grade 2 - 2.NBT.B.5

Each game includes a blank template so you can add new math problems.

3 Explore mathematics in nature

The world of nature is full of surprising examples of mathematics. The way things look, grow, or behave do so according to mathematical and scientific laws. Take your students outside and give them an enriching lesson about the world around them. Activities can include:

- Counting in nature, using rocks, leaves, or nuts as manipulatives (Common Core Math Standards: K.CC.A.1, K.CC.A.2, K.CC.B.4, 1.NBT.A.1)

- Discovering shapes that can be found naturally in nature, and outdoor shapes that are man-made (the spherical sun; a smooth, oval stone; a flat, round manhole cover; a rectangular walkway tile) (Common Core Math Standards: K.G.B.4, K.G.B.5, 1.G.A.1)

- Finding natural groups of objects in nature, such as groups of rocks, mounds of sand, or piles of leaves, to discuss the concepts of few, some, many, which item is bigger, which item is smaller, etc. (Common Core Math Standards: K.MD.A.2, K.MD.B.3, 1.MD.A.1, 1.MD.A.2, 2.MD.A.4)

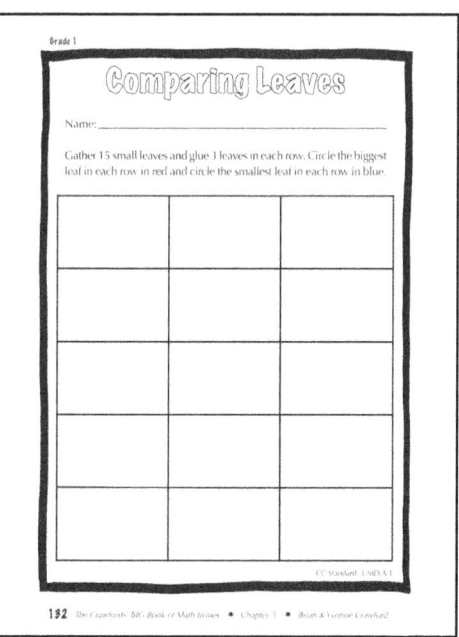

- Exploring the concept of time in nature, by finding out at what time of the day a flower opens up, at what time of day it is the hottest outside, etc. (Common Core Math Standards: 1.MD.B.3, 2.MD.C.7)

The activities Counting Leaves (kindergarten), Comparing Leaves (grade 1), and Estimating Leaf Length (grade 2) are included, beginning on **page 145**. Each activity aligns with Common Core Standards for those grades.

4 Measure outdoors

Ideas for activities include:

- Measurement: measure trees, distances, people, flowers, and rocks (Common Core Math Standards: K.MD.A.1, 1.MD.A.1, 1.MD.A.2, 2.MD.A.1, 2.MD.A.2, 2.MD.A.3, 2.MD.A.4).
- Statistics: measure the heights of plants and make a graph to illustrate your findings (Common Core Math Standards: 2.MD.D.9, 2.MD.D.10).
- Estimation: measure the height of a tall tree to a reachable point, then guess how many more equal measurements it would take to equal the height of the tree (Common Core Math Standard: 2.MD.A.3).

TIP

You don't have to measure with just the standard rulers or meter sticks. It's also fun to measure using objects or people, or to compare the heights of different objects.

5 Work in outdoor math centers

Set up temporary math centers outdoors to correspond with the curriculum or Common Core Standards you are teaching. For example, students at one center could be learning about measurement and volume using buckets of sand in a sandbox or by finding leaves and measuring their dimensions. A second group could be learning about time and distance by finding out how long it takes them to run from one tree to another, and a third group could practice adding all the kids on the soccer field or the clouds in the sky.

6 Plant a garden

Gardening offers tremendous opportunities to learn about math and science while creating something you and your students can be proud of. Gardens are long-term projects requiring care and upkeep, but the rewards are well worth the hard work.

Opportunities for practicing math skills while gardening include:

- Creating Mathbook prompts that directly correlate with the garden you are planting. For example, students can add or subtract seeds and measure plant sizes over time.
- Keeping track of how many seeds you're planting, how far apart you're spacing them, and how deep you need to dig the holes to plant them.
- Estimating how tall you think plants will eventually grow.
- Comparing different plants. Which plant grows the fastest? Which plant is the largest? In what other ways can you compare one plant to another?
- Classifying plants into different categories. Are there more white flowers, pink flowers, or blue flowers?

TIP

If you would like to learn more about gardening with your students, check these helpful resources:

www.schoolgardenwizard.org
www.kidsgardening.org/school-gardening
www.agclassroom.org/
bnan.wikispaces.com/Garden+Activities

Kindergarten — Outdoor Mathbooking Prompts

1. How many trees can you find in the playground? Count them!

CC Standard: K.CC.B.5

2. How many steps does it take to cross the baseball diamond?

CC Standard: K.CC.A.1

Compare 2 or more plants. Which plant is growing faster? Which plant is the tallest?

③

CC Standard: K.MD.A.1

Look at the garden. Can you group the flowers into categories? Draw pictures to help you decide how to sort the flowers. How many of each type of flower do you see?

④

CC Standard: K.MD.B.3

Look at your watch. What time is it? Now run across the soccer field. What time is it now? How many minutes did it take you to run from one end of the soccer field to the other end?

1

CC Standard: 1.MD.B.3

Look at the school. Subtract the number of doors you see from the number of windows. Draw pictures to help you solve this problem.

2

CC Standard: 1.OA.C.6

How many cars are in the parking lot? Without counting the number of tires on each car, can you figure out how many tires there are in total? Draw pictures to help you solve this problem.

CC Standard: 1.OA.C.6

③

Count all of the flowers you can find up to 120. Write that number in words.

CC Standard: 1.NBT.A.1

④

Find some leaves that have fallen to the ground. Count the leaves by 5s. Now, count the leaves by 10s.

1

CC Standars: 2.NBT.A.2

Find a fall leaf and paste it in your journal. Now measure the leaf. How tall is it? How wide is it? Measure it in both in centimeters and in inches.

2

CC Standars: 2.MD.A.1, 2.NBT.A.2

Plant some seeds. Measure how far you are spacing the seeds. Also, measure how deep you are digging the holes for the seeds.

3

CC Standard: 2.MD.A.1

Find a plant and estimate its height. Now measure it. Compare this plant to another plant. Which plant is growing faster?

4

CC Standards: 2.MD.A.1, 2.MD.A.3

Name: _____

Hide, Seek, and Solve
Tally Sheet

Problem	Answer	Problem	Answer
1		10	
2		11	
3		12	
4		13	
5		14	
6		15	
7		16	
8		17	
9		18	

Hide, Seek, and Solve — Kindergarten

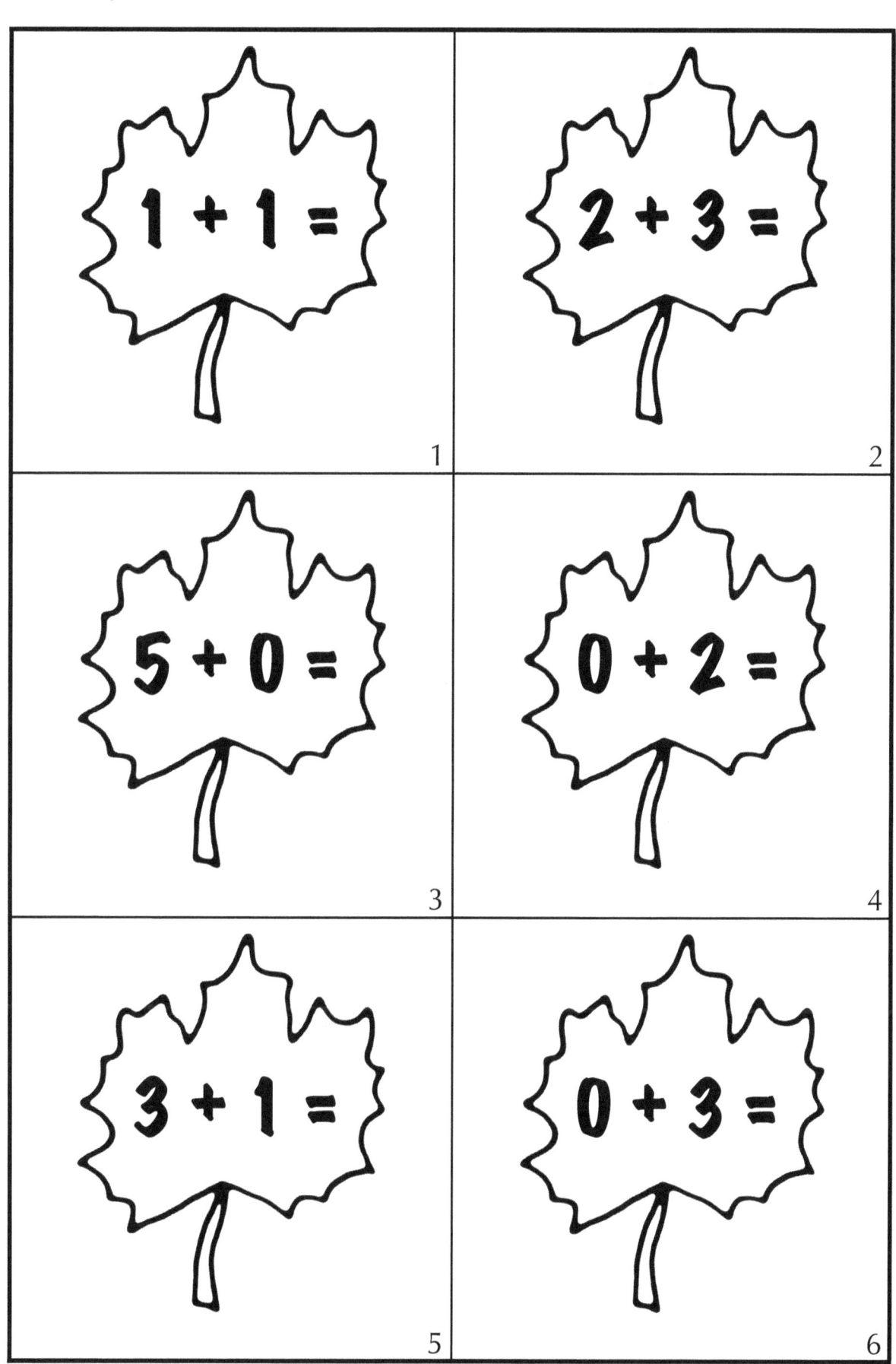

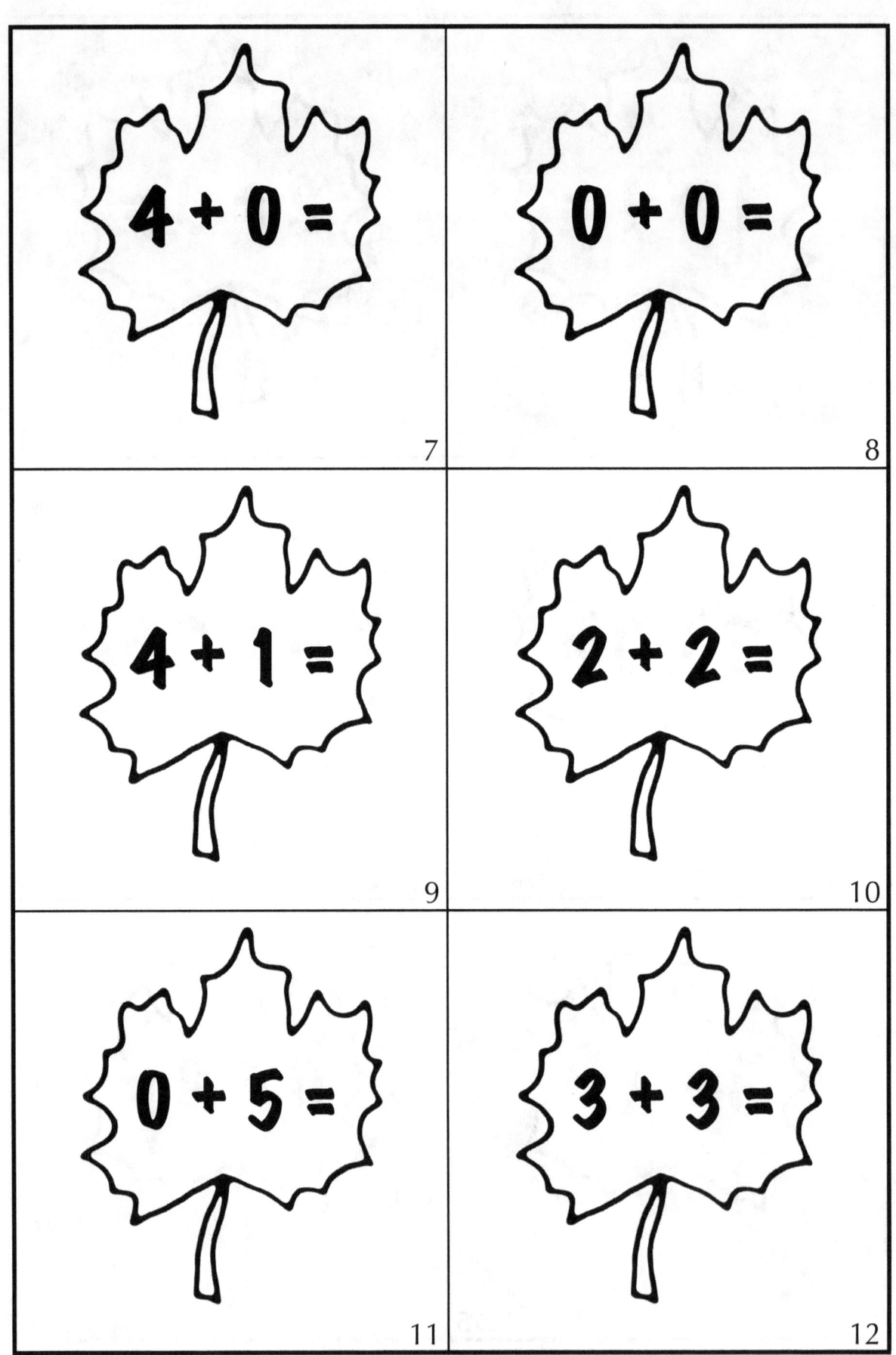

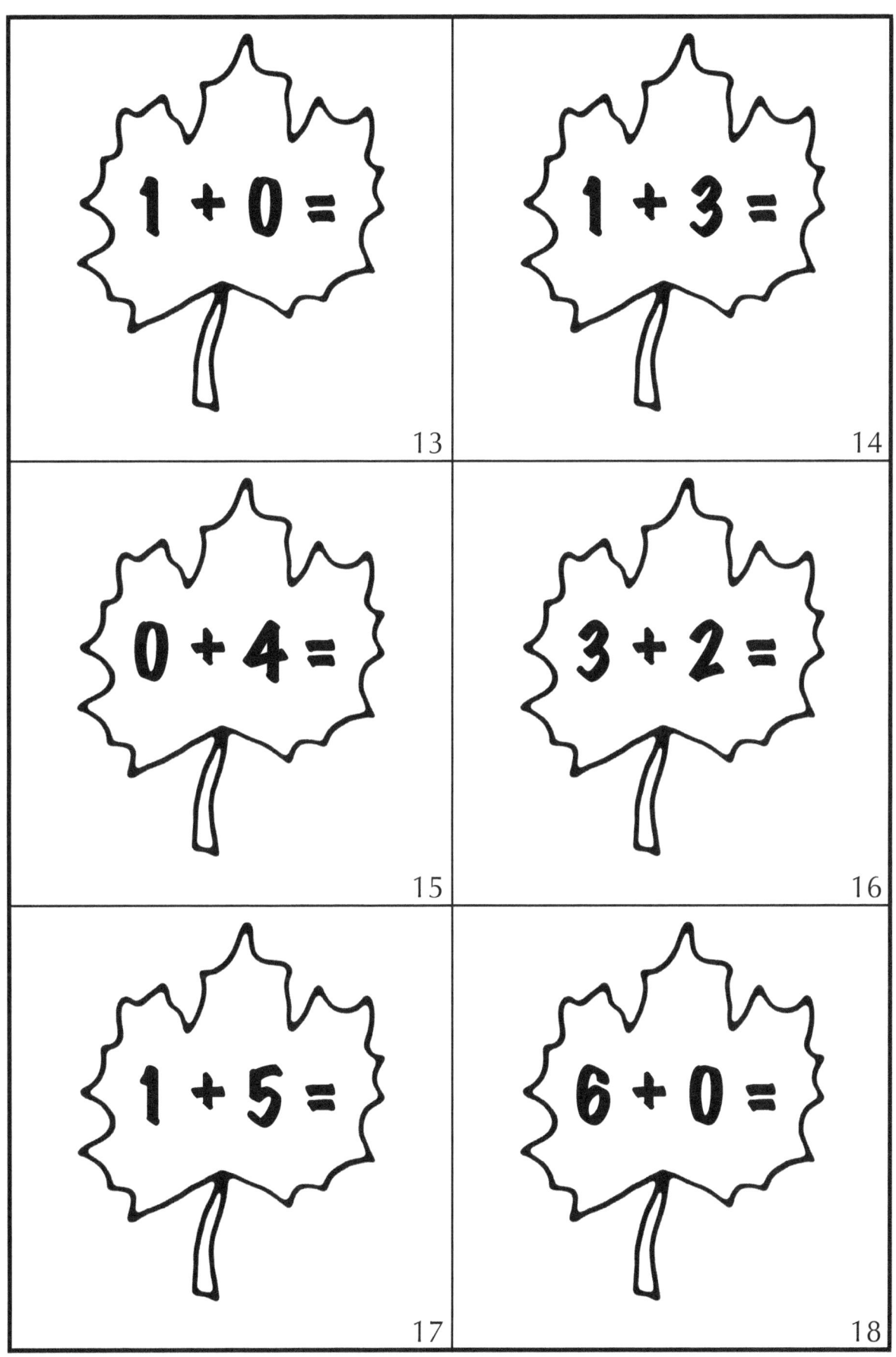

Kindergarten — Hide, Seek, and Solve

Hide, Seek, and Solve — Grade 1

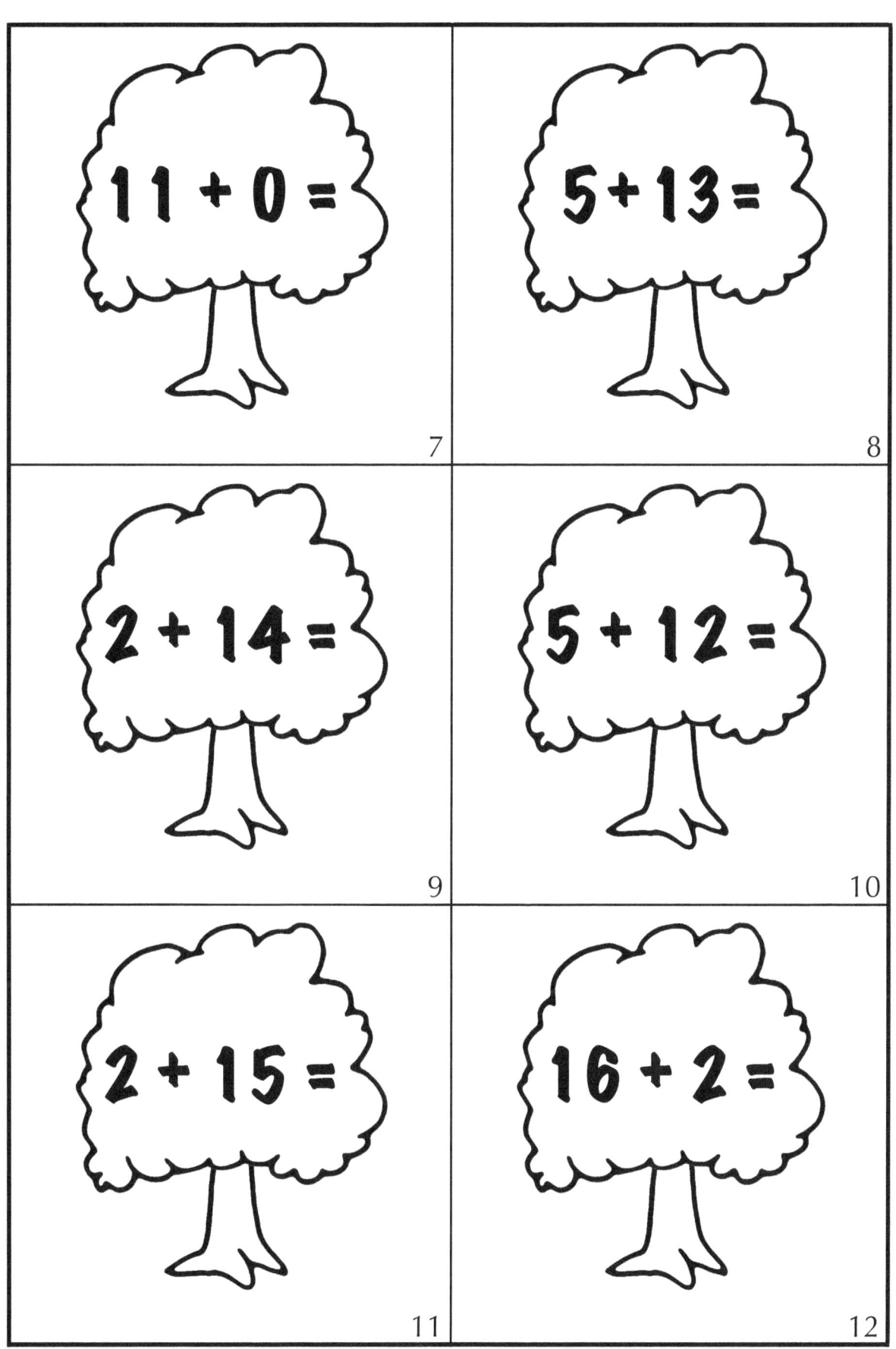

Hide, Seek, and Solve Grade 2

Brian & Yvonne Crawford   *   **Chapter 5**   *   The Crawfords' BIG Book of Math-tivities

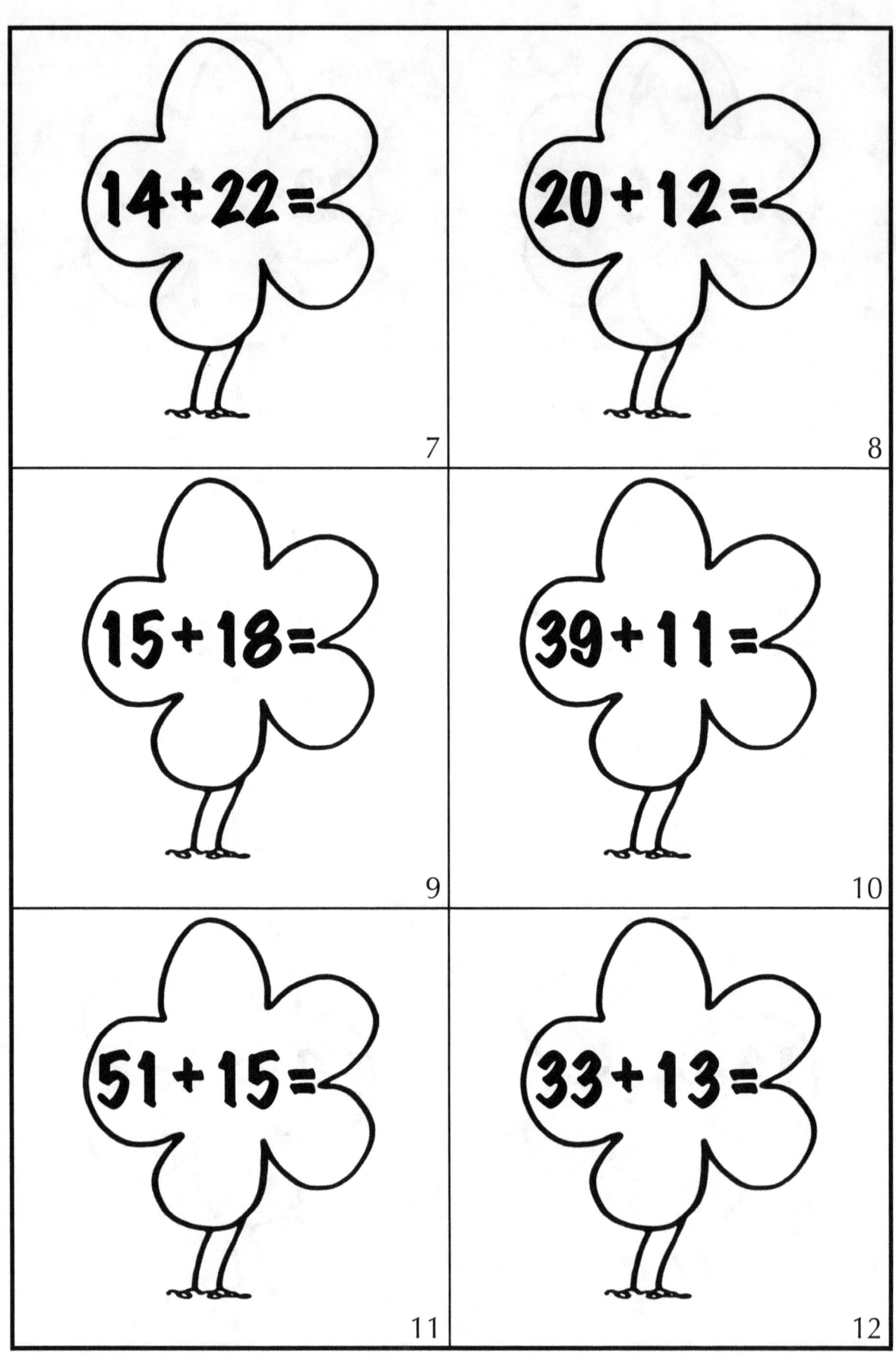

Hide, Seek, and Solve — Grade 2

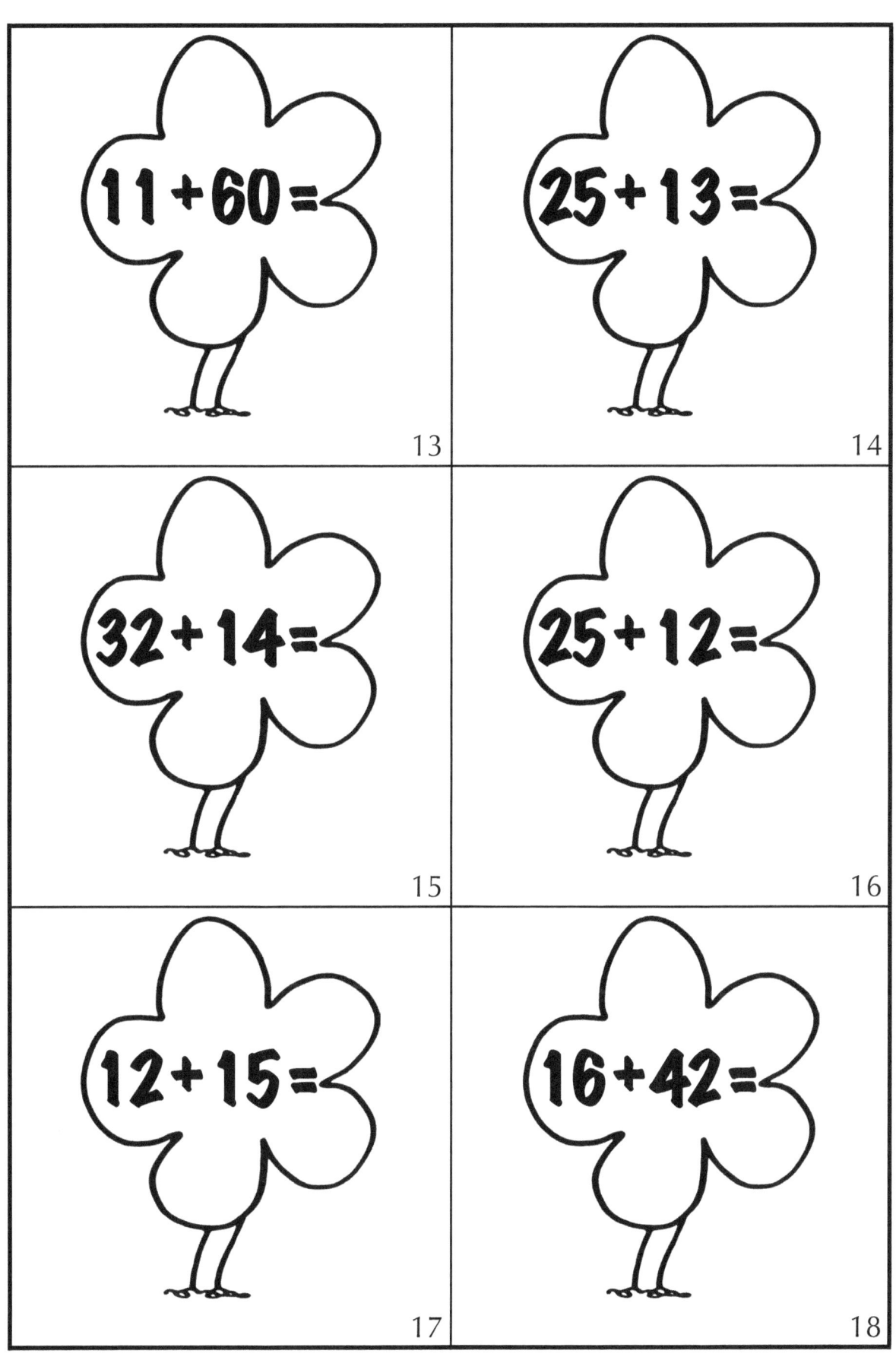

Counting Leaves

Name: _____

Find small leaves on the ground and sort them into groups of similar size and color. Then glue the leaves to this page. After you are finished, count the total number of leaves you have found.

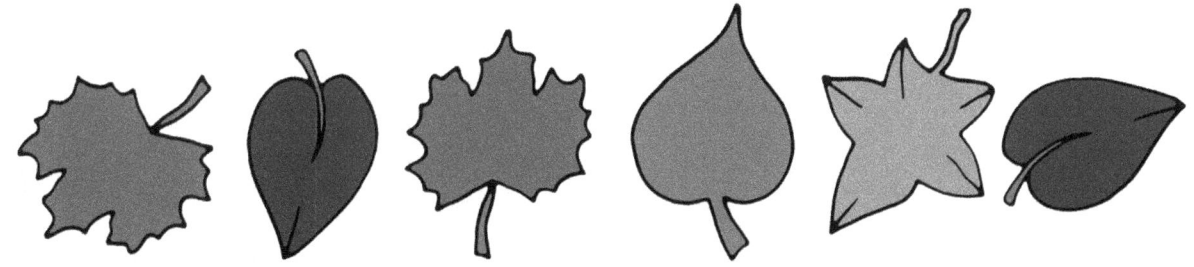

CC Standards: K.CC.A.1, K.CC.B.4, K.MD.B.3

Comparing Leaves

Name: _____

Gather 15 small leaves and glue 3 leaves in each row. Circle the biggest leaf in each row in red and circle the smallest leaf in each row in blue.

CC Standard: 1.MD.A.1

Estimating Leaf Length

Name: _____

Gather 5 small leaves and glue 1 leaf in each box in the **Leaf** column. Estimate the length of each leaf in centimeters or inches. Then use your ruler to measure each leaf.

Leaf	Estimate	Measurement

CC Standard: 2.MD.A.3

Chapter 6

Holiday and Seasonal Math

Take advantage of the calendar by incorporating holidays, current events, and the seasons of the year into your math lessons. Tailoring your math lessons to what's happening day-to-day keeps the learning fun and timely!

Incorporating the upcoming holiday can be very simple. If the kids are excited about Thanksgiving, have them count Pilgrim hats instead of books, or turkeys instead of buttons.

Besides the usual holidays, there are "school holidays" that you can celebrate: i.e., back to school (the first day/week of school), the hundredth day of school, the birthday of the person who the school is named after, and so on.

In addition, there are plenty of local special events you can celebrate: homecoming, a big game for your local baseball team, the anniversary of when your school was founded, or a fall or spring festival. Use your imagination and you'll find holidays all the time!

Incorporating Holidays

1 Use manipulatives and props that fit the season
Bring holiday-themed and seasonal items to your classroom to use as teaching tools. Before Halloween, for example, use small pumpkins to demonstrate halves, thirds, and fourths, or use plastic spiders for grouping manipulatives. If you're teaching addition in the wintertime, go outside to make—and add—snowballs. And if Easter is approaching, why not find out how many eggs you can fit into one basket?

TIP
Bring unique and interesting seasonal props to show your class, like dried Indian corn or spring flowers. Use these props to demonstrate mathematical concepts like counting, addition, and subtraction.

2 Create or alter worksheets
Create your own activities and games or alter the graphics and math problems on pre-existing activities to add some holiday flavor. For example, if you're using a standard textbook to teach math, it may include examples using objects such as apples, cupcakes, or tricycles. Take the numbers from those math problems and use those same problems, but with holiday-themed items instead: St. Patrick's Day shamrocks, New Year's Day fireworks, or groundhogs on February 2! The math skills being learned will remain the same, but the use of holiday-themed items to illustrate the concepts will make them more enjoyable and interesting to learn.

3 Reward your students with holiday and seasonal stickers and stamps
If you're funding your own rewards or are in charge of your own classroom budget, stock up on holiday and seasonal stickers and stamps when they are discounted.

4 Integrate your math and language arts curriculum
Take advantage of opportunities to integrate your curriculum. Tailor your math lessons around stories or books you are reading for English language arts. These are often themed to holidays or seasons of the year.

For example, if you're reading a book near Thanksgiving that features a family of Pilgrims coming to North America, you can create math problems using the same characters. Math problems might ask students how many turkeys are on the characters' farm, or how many hours the family spent bringing in the harvest.

5 Incorporate real-life situations into your math problems
If your students are wearing Halloween costumes to school, pose questions about the costumes themselves. How many students are dressed as witches? How many students don't have a mask? What percentage of the class is wearing a hat?

Respecting Cultures, Religions, and Traditions

While it's fun to incorporate holidays into your lessons, be careful to respect the beliefs of all of the students in your class. Schools often have policies about how to handle religious holidays, so make sure you are familiar with yours.

Take the opportunity to teach your students about different cultures. In December, you can incorporate a variety of different holidays into your activities—Christmas, Hanukkah, and Kwanzaa, for example. You can incorporate holidays such as Ramadan, Sukkot, and the Chinese New Year into your math problems during the year, and in the process, teach students a few of the traditions of people from many backgrounds.

In elementary school, some students will believe in Santa Claus and the Easter Bunny, but others will not. Students will likely turn to their teacher to settle the debate, so be ready to answer this question. You might choose to explain to your students that different people have different beliefs, and that these beliefs usually originate from our families or cultures. As such, the students' parents would be the right people to answer these questions. Be aware that giving a definitive answer or indicating that a family's beliefs are incorrect can easily land a teacher in trouble. Take care when answering them.

Seasonal Task Cards

Task cards are a fun activity that you can use in math centers in your classroom. Each task card includes an individual math problem to be solved, often accompanied by a fun diagram or picture related to the problem. Task cards are laminated and then bundled together in a booklet, using a single binder ring or a piece of string to hold them together.

TIP

Individual binder rings can be found in hobby stores or office supply stores. Use a single-hole punch to punch holes in the top right corner of each task card, then fasten the task cards together in a set using a binder ring.

Students work their way through one set of task cards and answer the questions on the cards on a separate sheet of paper. Once they've worked their way through the cards, they check their responses against the very last card that includes all the correct answers to the set of problems.

Sets of task cards usually address a single topic or incorporate a single theme. For example, one set of task cards can feature addition problems, and another set can feature subtraction problems.

Bring out season-specific task cards when those seasons arrive. When the season is over, store your seasonal task cards for next year's class.

How to create task cards:

- To make task cards, cut a piece of 8-1/2 X 11" heavy paper into four equal pieces, use the task card template in this book (**page 165**), or use blank file cards as a base for your task cards.

- Write one math problem on each card.
- Make sets of ten to twelve cards, and number each card in the set.
- Make one card with the answers to all the problems in the set so students can check their work, and put this card at the end of the set.
- Laminate the cards, and punch a hole in each one.
- Fasten the set together with a string or a large ring.

TIP

Color-code your task cards by Common Core Standard area for easy organization. For example, use a marker to add a green border to cards for Counting and Cardinality, a blue border to cards for Operations and Algebraic thinking, and a pink border to cards for Number and Operations in Base Ten.

Seasonal Task Card Sets

Beginning on **page 153**, sets of seasonal task cards for grades K - 2 are included, as well as the corresponding Common Core Standard to which each applies. Print or copy these onto heavy paper and cut them out to create your own classroom sets. Laminating the cards will extend the life of the set. A task card template is also included (**page 165**), for you to create new sets of task cards for future holidays and seasons.

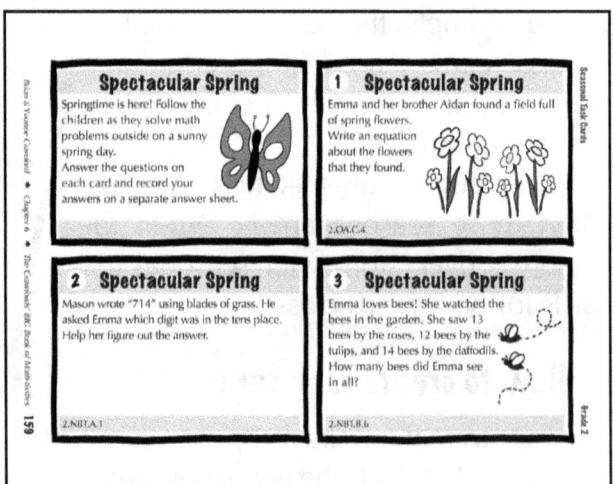

Seasonal Task Cards — Kindergarten

Harvest Time

It's autumn, and time to harvest the crops. Join these children as they harvest vegetables and help them solve their math problems. Answer the questions on each card and record your answers on a separate answer sheet.

1. Harvest Time

Lily started the day by harvesting tomatoes with her friend Natalie. Lily found a half of a tomato. If she has two halves of a tomato, what shape does it make when the halves are put together? Draw it.

K.G.B.5

2. Harvest Time

Natalie dug lots of carrots in the garden. How many carrots did she dig?

Groups of ten	Ones

K.NBT.A.1

3. Harvest Time

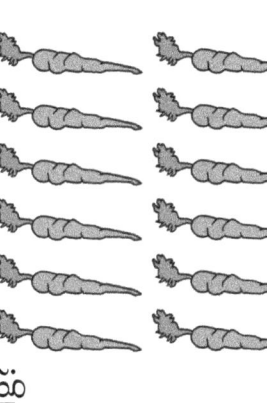

As Lily harvested vegetables, she found a rock. The rock had 4 equal sides. What shape was the rock?

K.G.B.4

Kindergarten — Seasonal Task Cards

5 — Harvest Time

Lily gathered 9 ears of corn and showed them to Farmer Ted. He asked her to write two addition equations that equal 9. Can you help her?

K.OA.A.3

7 — Harvest Time

Natalie is trying to figure out which basket will hold the most ears of corn. Which basket do you think will hold more corn, A or B?

K.MD.A.2

4 — Harvest Time

Natalie saw a hole in the ground. She drew the shape of the hole. Is the shape she drew a two-dimensional shape or a three-dimensional shape?

K.G.A.3

6 — Harvest Time

Farmer Ted said the kids could take a break and have some lemonade, but first, he asked them to answer this problem:

What number added to 8 equals 10?

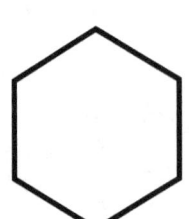

K.OA.A.4

154 *The Crawfords' BIG Book of Math-tivities* ✷ **Chapter 6** ✷ Brian & Yvonne Crawford

Seasonal Task Cards — Kindergarten

9 Harvest Time

One of Natalie's final chores for the day is to put away the supplies. Draw a basket beside the rake on the answer sheet. Draw gloves next to the rake. Draw a hat above the rake.

K.G.A.1

Harvest Time

Answers

1 ◯
2 1 ten; 2 ones
3 a square
4 a two-dimensional shape
5 $8 + 1 = 9$ and more
6 2
7 A
8 $3 + 5 = 8$
9 drawings will vary
10 ☐

8 Harvest Time

Lily harvested 3 potatoes and stopped to talk to Farmer Ted. Then she harvested 5 more potatoes. How many potatoes did she harvest in all?

___ + ___ = ___

K.OA.A.2

10 Harvest Time

Farmer Ted asked the kids to solve one more problem. He asked them to make a rectangle out of these two squares. Can you help them?

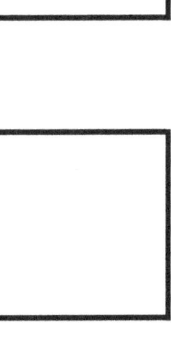

K.G.B.6

Kindergarten

Name: _____ Date: _____

Harvest Time Answers

1 _____ 6 _____

2 _____ 7 _____

3 _____ 8 _____

4 _____ 9 _____

5 _____ 10 _____

Seasonal Task Cards — Grade 1

Winter Wonderland

It's wintertime! Join a group of friends as they play in the snow. When they find math problems to solve, help them out! Answer the questions on each card and record your answers on a separate answer sheet.

1 Winter Wonderland

Sophia is making snowballs. She made 20 snowballs, but 10 of them melted. How many snowballs does she have left?

1.NBT.C.5; 1.NBT.C.6

2 Winter Wonderland

Sophia, David, and Jacob built this snowman. Then they drew a picture of it. Using paper clips to measure, about how many paper clips high is this picture of their snowman?

1.MD.A.2

3 Winter Wonderland

David and Sophia found a paper in the snow with two equations written on it. They want to figure out the relationship between the two equations. Can you help them?

$2 + 1 = 3$ $3 - 1 = 2$

1.OA.B.4

5. Winter Wonderland

David wrote this in the snow:

13 □ 2 = 11

Which sign should go in the box, − or +?

1.OA.D.7

7. Winter Wonderland

As Jacob was playing in the snow, he saw 8 + 12 snowflakes fall on his nose. Can you explain to him how he can use the properties of operations to solve this equation?

1.OA.B.3

4. Winter Wonderland

Jacob is counting snowmen in his neighborhood. He has already counted to 7. Help him count the snowmen. Start counting from 7 and end at 18. Just as he was he heading home, he saw 2 more snowmen. Count on from 18 to see how many snowmen he saw in all.

1.OA.C.5; 1.NBT.A.1

6. Winter Wonderland

Sophia's mom said that she would join in the fun if the kids could solve this problem. Help them solve it.

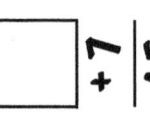

1.OA.D.8

Seasonal Task Cards — Grade 1

9 Winter Wonderland

When Jacob went home, his dad made him a cup of hot chocolate. He counted 12 marshmallows in the cup. He asked his dad to put 13 more marshmallows in his cup. How many marshmallows did he have in all?

1.NBT.C.4

Winter Wonderland

Answers

1. 10
2. 1 paper clip high
3. answers will vary
4. 7, 8, 9, 10, 11, 12, 13, 14, 15, 16, 17, 18, 19, 20
5. -
6. 8
7. answers will vary
8. 4
9. 25
10. ◯

8 Winter Wonderland

David saw many shapes in the snow. Can you count how many round shapes he saw?

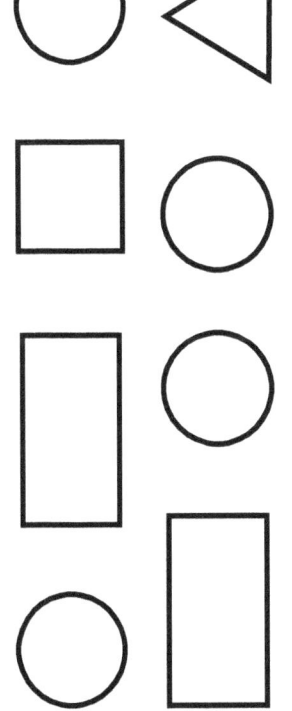

1.MD.C.4

10 Winter Wonderland

Jacob and Sophia saw two half-circles cut out of the ice on the frozen pond. Help them make one shape by combining them. Draw the shape on the answer sheet and name the shape.

1.G.A.2

Grade 1

Name: _____ Date: _____

Winter Wonderland Answers

1 _____   6 _____

2 _____   7 _____

3 _____   8 _____

4 _____   9 _____

5 _____   10 _____

The Crawfords' BIG Book of Math-tivities   *   Chapter 6   *   Brian & Yvonne Crawford

Spectacular Spring

Springtime is here! Follow the children as they solve math problems outside on a sunny spring day.
Answer the questions on each card and record your answers on a separate answer sheet.

1 Spectacular Spring

Emma and her brother Aidan found a field full of these spring flowers. Write an equation about the flowers that they found.

2.OA.C.4

2 Spectacular Spring

Mason wrote "714" using blades of grass. He asked Emma which digit was in the tens place. Help her figure out the answer.

2.NBT.A.1

3 Spectacular Spring

Emma loves bees! She watched the bees in the garden. She saw 13 bees by the roses, 12 bees by the tulips, and 14 bees by the daffodils. How many bees did Emma see in all?

2.NBT.B.6

5 Spectacular Spring

Aidan drew a circle in the dirt and asked his sister to divide it into 4 equal parts. Can you help her with this task?

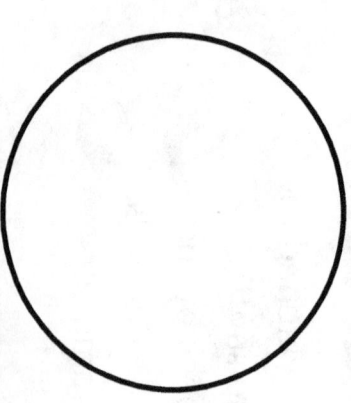

2.G.A.3

7 Spectacular Spring

Aidan planted 79 seeds in his garden, but he forgot to water 2 seeds. How many seeds did Aidan water? Use the number line below to help you solve the problem.

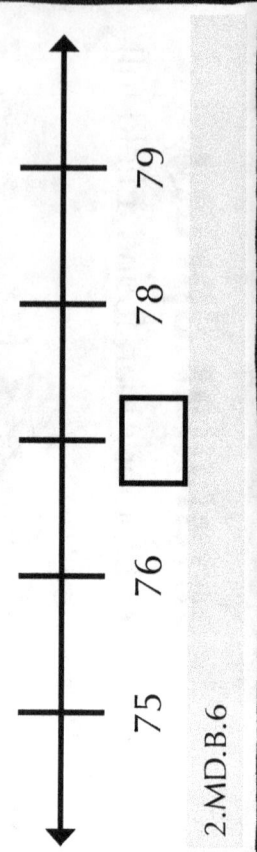

2.MD.B.6

4 Spectacular Spring

Mason found an amazing flower. Each petal had 3 equal sides and 3 equal angles! What shape are the petals of the flower that he found?

2.G.A.1

6 Spectacular Spring

Below is a chart of the different colors of flowers that Mason found. Which color of flower did he find the most?

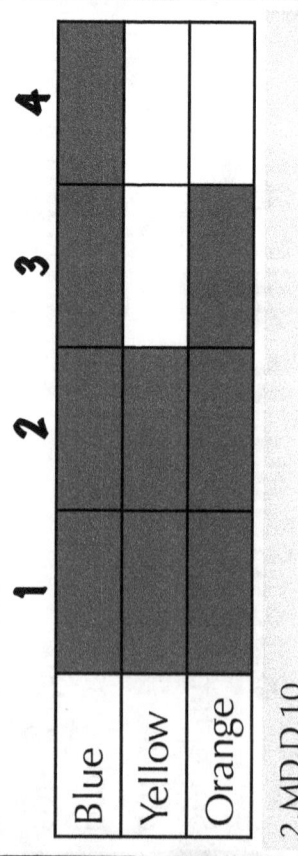

2.MD.D.10

Seasonal Task Cards — Grade 2

9 Spectacular Spring

According to this line plot, how many of Mason's rosebushes are taller than 3 feet?

```
                X
          X     X
          X     X   X
      X   X     X   X
  X   X   X  X  X   X
  2   3   4  5  6
  Rosebush Heights (in feet)
```

2.MD.D.9

Spectacular Spring

Answers

1 4+4=8 or 4×2=8   **6** blue

2 1   **7** 77

3 39   **8** 16

4 triangle   **9** 8

5 ⊕   **10** 366

8 Spectacular Spring

Aidan has divided his garden into squares. What is the area of his garden based on the number of squares?

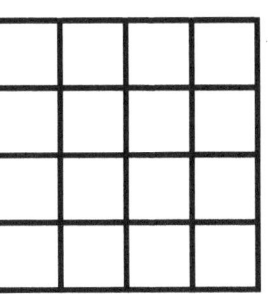

2.G.A.2

10 Spectacular Spring

Before Aidan and Emma went home, they decided to count the spots on the ladybugs they found. Aidan counted 125 ladybug spots and Emma counted 241 ladybug spots. How many spots did they count in all?

2.NBT.B.7

Grade 2

Name: _____ Date: _____

Spectacular Spring Answers

1 _____   6 _____

2 _____   7 _____

3 _____   8 _____

4 _____   9 _____

5 _____   10 _____

Notes

Notes

About the Authors

Brian and Yvonne Crawford are a husband-and-wife team with significant experience in elementary education, post-secondary education, online education, and home-schooling.

Yvonne has a BA in Linguistics, an MA in Sociology, and an MA in Teaching English as a Second Language. She spent several years teaching English and Language Arts in the United States, Hungary, Slovakia, and France. Brian has a BS in Business Administration, an internationally-focused American MBA and an MS in International Project Management from a French business school. He served as a faculty member and instructor at a top-ranked medical university in the United States and has several years of experience in online teaching and information sharing. Yvonne and Brian homeschooled their children for several years and made an effort to bring them up bilingual in English and French, spending several years abroad living and working in France, Spain, and Ireland.

Yvonne and Brian provide teaching tips and resources on their website, **Mixminder.com**.

About Compass Publishing/Brigantine Media

Compass Publishing is the educational books imprint of Brigantine Media. Materials created by real education practitioners are the hallmark of Compass Publishing. For more information, please contact:

Neil Raphel
Compass Publishing | 211 North Avenue | St. Johnsbury, Vermont | 05819
Phone: 802-751-8802
E-mail: **neil@brigantinemedia.com** | Website: **www.brigantinemedia.com**

Math + FUN = The Crawfords' BIG Book of Math-tivities!

Need some new ideas for teaching math to kindergarten, grade 1, or grade 2 students? Try some "Math-tivities" to enliven your math class and help your students learn.

★ **Mathbooking** — a new combination of math journals + scrapbooking

★ **Math Glyphs** — students answer math questions to create a personal work of art

★ **Math and Tell** — combines storytelling with math word problems

★ **Math Games and Puzzles** — brand-new math games to entertain students while they learn

★ **Taking Math Outside** — use the great outdoors as your new location for math class

★ **Holiday Math** — special seasons of the year are perfect for showcasing math

What's more, all the Common Core State Standards for mathematics in kindergarten, grade 1, and grade 2 are aligned with one or more "math-tivity," so you can be sure your curriculum is covered while your students are learning to love math!

"It's a dynamic way of approaching math."
JENNIFER AYERS, TEACHER, GRADE 2, CHATTANOOGA, TENNESSEE

"My students love how the lessons incorporate art and math."
MICHELLE DIVKEY, TEACHER, KINDERGARTEN, CALIFORNIA

"The teacher directions were very detailed and easy to follow."
FERN SMITH, TEACHER, GRADE 3, FLEMING ISLAND, FLORIDA

"The Crawfords make math crazy fun with their multisensory approach."
KIMBERLEY ARBOUW, TEACHER, KINDERGARTEN, GRADE 1, AND GRADE 3, ROANOKE, VIRGINIA

COMPASS
A DIVISION OF BRIGANTINE MEDIA
www.compasspublishing.org

ISBN: 978-1-9384062-9-4

$24.95

ABOUT THE AUTHORS

Brian and Yvonne Crawford are a husband-and-wife team who bring their unique experience to mathematics education.

Yvonne has spent several years teaching English and Language Arts in the United States, Hungary, Slovakia, and France. Brian has served as a faculty member at a top-ranked medical university in the United States and has a number of years experience in online teaching and information sharing. Yvonne and Brian homeschooled their children while they lived in several countries in Europe.

Yvonne and Brian provide teaching tips and resources on their website, Mixminder.com, where many of their concepts have been shared with their readers and adopted in classrooms everywhere.

www.ingramcontent.com/pod-product-compliance
Lightning Source LLC
Chambersburg PA
CBHW080248170426
43192CB00014BA/2598